# Fundamentals of Physical Geography

# Fundamentals of Physical Geography

### PART - 1
### LITHOSPHERE

**Prof. Girindra Kumar** (Retd.)

ISBN: 978-93-5324-268-8 (HB)

First Published, 2022

*Published by*

Kalpaz Publications
C-30, Satyawati Nagar,
Delhi – 110052
Ph.: 011-47034999, 9811692060
www.kalpazpublications.com
E-mail: kalpaz@hotmail.com

*Printed at:* G. Print Process, Delhi

# Contents

# Preface

Geography is concerned with the *explanation of intrinsically related* but *objectively definable phenomena* that find expressions *in the earth system*. Study of physical geography assumes greater significance when one realizes that no explanations related to the human world can be made without understanding the underlying processes that characterize our living planet. This, then clearly involves not only the information about its surface expressions but also of an understanding of the Earth's physical being (or part of it), its relation with other celestial bodies, and its changing endowments since its origin. In this regard, it must be noted as **Audouze** (1988: 12) observes that "every part of the Universe (the Earth included) appears to exert a direct influence on the rest of the Universe and in turn, similarly to be affected by the rest of the Universe". It was probably this understanding that **Newton** attempted to formulate a uniform **law of motion** on the Earth and in the intergalactic space. It was probably the same understanding that **Darwin** attempted to integrate the evolutionary processes of the cosmos with that of terrestrial organisms (**Christian** 2011:16).

It, therefore, makes it almost mandatory for students pursuing study of any subsystem of the Universe to understand the differentially interacting processes that has been shaping it with accordant resultants. In the light of the foregoing such studies may, therefore, be grouped under two categories namely:

(1) Study of the Universe as a hierarchical system, and
(2) Multiple forms and structures obtained on increasingly smaller scale.

As a matter of fact Physical Geography is supposed to synthesize the two categories for, the Earth as a system is not entirely a closed system. It very often is influenced by a much larger system of interstellar space and which physical geographers must try to understand in order to explain the characteristics within the earth system.

At the same time, as geography is primarily concerned with the analytical study of earth phenomena, physical geography as its sub branch must also not overlook the Earth as a **closed system**. This necessitates an understanding not only of the unique **land-atmosphere ocean system** that makes the planet earth different from other known planets but also of the processes through which they have been obtained and are made to operate differentially. It must, therefore, include the study of subsystems operating on earth almost as **open systems**. It is, thus, imperative that physical geographers study the physical properties of the Earth as a body that includes understanding of its structure and distribution of mutually linked surface as well as sub surface phenomena- climatic, oceanic and biogenic including man, and which impinge on each other. It is true that with the growth of human civilization and concomitant technological advancements, the control shackles of nature on earth system appear to have increasingly been loosened. Yet the space and spatial characteristics continue to exert substantial influence on the use of ever evolving technology and the human induced processes. However, growing intervention by man (knowingly or unknowingly) is found to have modified the natural processes (at least over and near the earth's surface) supposedly to improve the level of human living at the cost of nature's resilience. It is, however, being increasingly realized that developing global situation is detrimental to the survival of humanity- present and future.

Present book is an attempt to explain these basic linkages and their significance in earth system through the study. However, the present presentation is devoted only to the evolution and explanations of the Earth and its physical properties. An attempt has been made to integrate the changing views of scholars over space and time. It has been done also to impress upon this book users that any existing view(s) and interpretation(s) are the results of man's incessant quest for facts. The book is expected to enable its readers particularly the students to develop a rational vision and skill to interpret the mutually impacting phenomena that make the Earth unique in system of planets.

I am thankful to my erstwhile colleagues and students of Geography Department of Mizoram University who have always been encouraging me to indulge in such kind of endeavour. I am also grateful to my wife Gauri who saw to it that unnecessary family chores do not disturb my thought process while writing this book. I cannot ignore my two sons Gaurav and Gahan and their wives Rakhee and Shalini who despite having trainings in other disciplines went through this presentation and on specific places helped me to modify my language for common understanding. I am also thankful to my grandsons and granddaughters

Master Shaurya, Master Agustya, Miss Ishana and Miss Gayatri (all between 8 years and four years of age) who at every step of my writing wanted me to explain things and figures in their inquisitiveness. It helped me in modifying my language further so that any student with primary knowledge of English may understand the idea behind this book. It will be a crime if I do not acknowledge the use of figures and views of different writers and scholars in this book. Though I have tried to acknowledge their contribution at relevant places I apologize if some names have been inadvertently left out.

I will also be grateful to the readers and users of this book who bring to my notice the mistakes and misinterpretation/misrepresentation of facts and which may be rectified in future editions.

**Girindra Kumar**
(Retd. Professor, Mizoram University)
Visiting Professor Manipur University
gkumar_1951@yahoo.com

# List of Tables

# List of Tables

# Chapter - 1

# Universe and the Solar system

## Objectives of the Chapter

Main objective of this module is to make its readers understand the different dimensions of cosmogony and the interstellar relationships with particular emphasis on our solar system. The module undertakes the study in respect of

- Origin of the Universe
- Solar System and its position in Universe
- Shape, size, distances and motions of planets.
- Sun and its Energy
- Relationship with human world

## Chapter - 1:1 Origin of the Universe

Scholars since the beginning of the human civilization have been having a keen interest in developing a unified knowledge that could explain the phenomena both visible and invisible. But widening and deeper understanding of different aspects over time has given rise to various explanations and sometimes contradictory opinions in different fields of knowledge.

Opinions about the origin of the Universe likewise also have been diverse. As a consequence of ever developing technologies and information obtained thereof they are also found to have undergone substantial changes since their initial postulates. The scholar, however, seem to have convergence of opinion about the definition of Universe. The Universe, to all of them is the representative of **'everything that exists'**. It includes all matters- smallest particles to the largest ones irrespective of their visibility; and energy reflected through their **luminosity** or absence of it (**dark matters**). In other words, the Universe includes everything that human mind can think of. How they came to exist continues to be a vexed question even today. And so does the question of origin of the Universe.

Based on available information a number of theories have been propagated at different periods of human history. They, however, may broadly be categorized in two major groups as (i) **Steady State** that may include **Creation myths** and (ii) **Big Bang (Evolutionary)** theories.

## Steady State theory

Steady State theories in one form or the other dominated the discourse on origin of the Universe in one way or the other since the beginning of the human civilization till early 17th century A.D.

The theories under this group have been developed around the premise that the Universe is **eternal** with no beginning and no end. They, in general, envisage that the input, output and properties of a system remain constant over time. Significance of steady state group of theories lies in the fact that they provided a scientific base for simplification of complex natural system of the Universe. Initially it found expressions in 'creation myths' advanced by many old scriptures across the world. It, basically, was built around the premise that '**Time**', an essential component of creation myths, is without any boundary and is limitless. However, in its linear progression it has been bound to breed difference in matters. Increasing differentiation over time continually has been breeding the growth of opposite properties in them causing what **Christian** (2011:19) calls '**fundamental clash of opposites**'. This clash begets more difference with varying interactions reflected in multitude of phenomena over time and space.

The Universe, under this theory, though is considered to represent growing diversity; relative spatial position of the original cosmic bodies is believed to be fixed. It was probably this understanding of the cosmos that led **Aristotle** (4th B.C.) and later **Ptolemy** (first century A.D) to place the Earth in the centre of the Universe. Around this premise Ptolemy developed a model of cosmos in which the Earth was encircled by eight spheres. These spheres were represented respectively by the moon, the Sun, Mercury, Venus, Mars, Jupiter, Saturn (then known planets) and the outermost sphere of **fixed stars** which moved around the Earth. The nomenclature 'fixed stars' probably was derived from the fact that they appeared to maintain their same relative position with respect to each other (*not necessarily in relation to the Earth*). Despite erroneous assumption, it enabled ancient astronomers to predict the positions of the celestial bodies and cosmic events like eclipse, alignment of planets and stars etc. with reasonable precision. The idea though not universally accepted even then found much support from the **Judean-Christian-Islamic** scriptures. They, particularly the Christian Church in post **Crusade** period, considered the view in tune with the idea of creation till early 17th century. The basis of such a support as **Hawking** (2011:4) believes was founded in the fact that it 'left lots of room outside the sphere of fixed stars for heaven and hell' which regulated **social behaviour** of most of the societies.

Ptolemy's model came under severe criticism when **Copernicus**, himself a Polish theologian, put forward his model (under a pseudonym) of the solar system in 1514 A.D. Based on his observation and calculation of the apparent motion of the planetary bodies he opined that the Sun and not the Earth was stationary and that all the planets including the earth moved in *circular orbit* around it. This view despite some flaws found greater acceptance only after a century when two scholars **Johannes Kepler,** a German and **Galileo Galilei,** an Italian supported the idea with some hard evidences. Kepler modified Copernicus' view with the proof that all the planets including the Earth revolved round the Sun in an *elliptical orbit* rather than in a circular orbit. It implied that Ptolemic models were rejected by 18th century. Invention of **telescope** in 1609 simultaneously enabled Galileo to observe planetary and satellite motions. He found that the **Satellites of Jupiter** moved around it and not directly around the Sun. He also concluded that heavenly bodies did not directly revolve round the Earth. Their paths were much more complicated than believed by the scholars of Aristotle-Ptolemy tradition.

By and large invention of telescope and formulation of laws in the field of geophysical science (**law of gravity, laws of motion, law of relativity** etc.) and development of associated technology since, enabled astrophysicists to observe the interstellar space more closely and accurately. They realized that the Universe was much bigger than commonly believed and it was continually expanding. The findings provided the basis for presently most accepted **theory of Big Bang** glimpse of which is found in **Emmanuel Kant's Nebular hypothesis**. However, the findings supporting the Big Bang were criticised by 20th century pioneers of steady state theory led by **Herman Bondy, Thomas Gold** and **Alfred Hoyle** in 1948. Their proposition was different from the earlier creation theories. They attempted to integrate newly found facts about the Universe with that of Creation theories. They suggested that the galaxies of the primordial Universe have been moving away from each other within its *finite dimension*. In the process of its spatial expansion voids were created regularly in between galaxies. It facilitated creation of new matters continually in those voids. Consolidation of new matters formed new galaxies which filled the **intergalactic space**. The Universe, therefore, would broadly appear similar over time and from any point in space despite increasing intergalactic distances. It, therefore, implied that despite variability in spatial dimension relative positions of cosmic bodies would continue to be almost unchanged.

The Steady State theory of 1940s, however, could not hold for long as new discoveries since could not be satisfactorily explained by it. As a result, there are very few scholars who support the theory at present.

## Merits

- The theory is constructed around simple traditions of different religious beliefs and is legible to common mass of people.a.
- It advocates creation of the Universe at a finite and easily comprehensible past. St. Augustine (16th century) based on the **book of Genesis** placed the age of the Universe around 5000 B.C.
- Proponents attribute the creation to some cause that makes it easy to explain the existence of the Universe.
- It suggests creation of man and the creation of Universe to be simultaneous giving credit to **divine intervention**.
- Formation of new matters and galaxies could be scientifically explained in a **finite universe**.

## Criticism

The steady state theory as proposed by Bondi, Gold and Hoyle though is considered by many to be 'a simple and good scientific theory' has been criticised on the basis of many earlier and later findings.

- Newton had observed that if the universe were finite then **gravity** would have drawnall the matter into the centre of the universe. It suggests that intergalactic phenomena could not be formed in the manner suggested by Bondi et al.
- **Studies in thermodynamics** reveal that the functional energy of the Universe has continuously been decreasing. It implies that a time will come when no energy will be left to create anything

as envisaged under steady state theory. This means that there was a time when there was more energy available to create things of diverse nature. In other words, the Universe had a beginning when energy was created and it is likely to exhaust sometime in future. This is against the proposition under steady state theory that assumes a universe without beginning and so without an end.

- **Edwin Hubble**, an American astronomer in 1929 established that the distant stars were moving from each other and spatial dimension of the Universe was increasing. It also implied that inner part of the Universe was balanced by mutual gravity of the stellar bodies whereas outer part was expanding due to forces opposed to gravity i.e. **antigravity forces** for, gravity is considered to be a function not only of mass but also of distance (*gravity is inversely proportional to increasing distance*).

- The steady state theory 'required a modification of general relativity to allow for the continual creation of matter to maintain density of the Universe. But the studies on radio wave interceptions suggest that the Universe must have been denser earlier. The findings obviously contradict the basic tenets of the steady state theory.

## Big Bang

Big Bang theory started being constructed since the beginning of 19th century against the steady state theories of eternal universe. It may be noted that even **Einstein** under theory **of general relativity** developed in 1915, appeared to be favouring a static Universe. However, **Alexander Friedman**, a Russian scientist, combining his two simple assumptions with that of law of general relativity was able to prove theoretically in 1922 that the Universe was in fact not static but expanding. His simple assumptions were- (i) Universe looked identical in any direction from a point and (ii) it would look identical from any point in the universe. Most of the scientists since have been moving towards an evolutionary theory. Later it was confirmed by **Edwin Hubble** in 1929 who catalogued many other galaxies and their increasing distances based on **stellar spectrum.** In other words, he confirmed that the Universe was expanding. It logically meant that in some fathomable past different galaxies must have been closer. The finding provided a solid ground to establish the theory popularly known as Big Bang. More evidences were obtained from the analyses of other aspects of the interstellar space like **intergalactic sources of radio**

**waves** (Cambridge University Project, 1950-60) and studies on **microwave radiation (Penzias and Wilson**, 1965) etc. suggesting that the Universe has been experiencing change in its density from high to low and it was expanding at a rate of about 258 million kilometres per second (about 90 percent of the speed of light) that defied forces of gravity. Finding the rate of expansion scientists have been able to find out the time when the Universe might have started to expand. All the findings appear to suggest that the Universe started to expand about **13.7 billion years ago** (though Hubble's initial calculation suggested a time only of 2 billion years i.e. less than the origin of our solar system and was modified by later geophysicists to be between 10 and 20 billion years). The expansion, it is believed, was triggered due to accretion of infinitely small, energetic and dense burning matters with **unfathomable density and space- time curve'** and which were compressed to the size probably of nucleus of an atom at zero interspatial distance. They assign the term 'big bang to the event. The incidence, most of the scientists believe, marks the **beginning of space, time, matter and energy**. What led the earlier chain of events leading to formation of energy and matter which condensed to form the primordial energetic and dense matter is difficult to explain despite the fact that quantum physics suggests that matter may emerge from apparent nothingness.

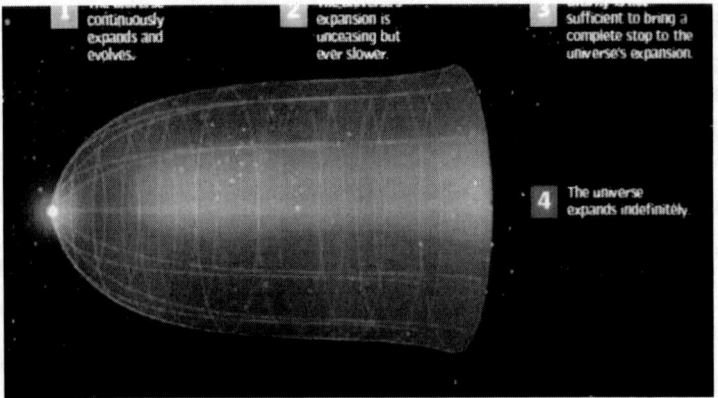

*Figure 1.1: The Big Bang from apparent nothingness*

The theorists of Big bang assume that the primordial subatomic particles interacted with each other due to very high levels of radiation to form atomic nuclei. Nuclei's high density introduced matter and intense cosmic inflation which continues to expand the universe. The universal space initially, it is believed, was occupied by a cloud of uniformly dispersed elemental **hydrogen and helium gases**. Subsequently the gas particles under varying temperature conditions started to organize themselves in innumerable network of **galactic filaments** with voids in between. Immeasurable space occupied by the ever expanding Universe, it is now established, is interspersed with clusters of galaxies (visible matters and energy) and **black holes** (dark matters). These matters, it is now almost accepted, also form the founding substances of the biotic world. The scientists have attempted to trace the evolution of the Universe in which our solar system came into existence almost 900 billion years after the Big Bang. The following timeline outlines the stages in the growth of the Universe.

**Stages in Evolution of the universe**

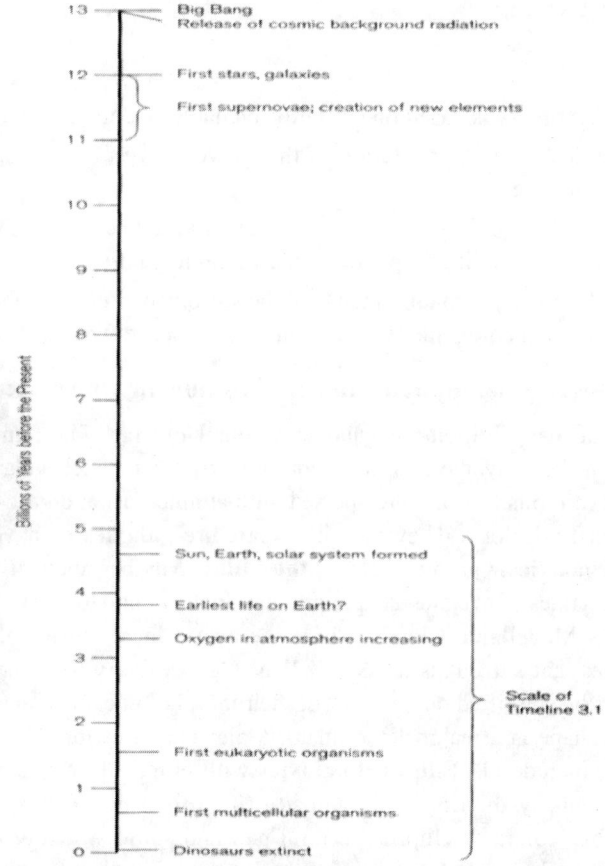

*Figure 1.2 : Time line (Adopted from Christian, 2011:45)*

Despite ambiguity of the conditions before the Big Bang, modern science since has been able to explain the evolution of the Universe with coherent logic. The cosmologists now are fairly certain that within first 300, 000 years of the Big Bang there was a period when initial subatomic particles organized themselves into atoms of immense energy and matter of immeasurable density in a nonexistent space till it could not contain any more. It, therefore, had to explode. Thus, 'basic particles of the material Universe necessary for creation' became available to the present space of the Universe. These basic particles since are believed to be organizing themselves in different patterns under different temperature and temporal dimensions. All that makes our present Universe is the reflection of recurrent arrangements and rearrangements of the same energy and matter that was available some 13.7 billion years ago in a nonexistent space.

## Merits

- The theory is based on currently available scientific information.
- It describes the evolution of the Universe with coherent logical sequence.
- It, at least in its initial formulation, is similar to creation myth and which made it acceptable to the Catholic Church in 1951.
- The theory explains many of the unsolved problems faced by earlier theories like why the interstellar space was not bright?*

### Chapter-1:2: Solar System And Its Position In Universe:

There exist many billions of galaxies in our Universe. The number of galaxies in the universe is still not definitely known. Each galaxy is a cluster of millions of stars interspersed with atomic dust and gases. They do not exist all alone. They may have **satellite galaxies** or have **local group of galaxies**. Our own galaxy- **the Milky Way** has about fifty nine satellite galaxies of irregular shapes gravitationally bound to it. The largest of them is Magellanic Clouds. There are also a local cluster of about 15 galaxies that surrounds the Milky Way. Galaxies vary in shape. They may, thus, be classified on the basis of their morphological characteristics of which shape is a major determinant. Major classification of galaxies, therefore, include (1) **Elliptical galaxies** with elliptical or round shape having regularly distributed luminosity (2) **Spiral galaxies** with two components- round or elliptical central part and spiral structure beyond on the same plane, (3) **Lenticular galaxies** though with a central bulge and disc on the same plane but without spiral structure, and (4) **Irregular**

**galaxies** having no regular structure. Our galaxy- the Milky Way with a radius of about 19500 light years, is spiral in shape.

(Mark the uneven distribution of matters that make our galaxy)

*Figure1.3: Milky Way the Spiral Galaxy.*

Our Solar system is located almost at the outer skirts of its central spiral. The distance of our solar system from the centre of the Milky Way is about 10400 light years.

Our Solar system is only a tiny part of the Milky Way galaxy and our sun is only one of the many billions of stars in it. Many stars like our sun are believed to have a system of different number of [1] (German philosopher **Heinrich Olbers** had suggested that in an infinite static universe every line of light from distant stars would end up in straight line on the surface of other stars making the universe all the time as bright as the surface of the Sun and without the darkness of night. The concept is popularly known as **Oblars' paradox**.) planets bound to them by their gravitational force. Planets may also have one or more satellites. The satellites are bound to their parent planets by gravity and revolve round them.

## Chapter - 1:2 (A): Solar System

The Sun is at the centre of a well-structured system known as solar system. The system consists of eight planets with their satellites*. They have been categorised in order of their distances from the sun as well as

**Figure 1.4: The sun and planets in their proper order and
approximate proportion of their size and distance from the sun.**

*\*Pluto considered being the ninth planet till 2006 is not considered a solar planet
under the new definition of planets by International Astronomical Union. Instead its
solid icy body is considered to have been captured by the gravitational pull of our Sun
from a disc shaped region of the Kuiper belt that exists in the neighbourhood of
Neptune. Unlike rest of the outer planets Pluto is almost the size of our earth and
cannot be considered under the group of giant planets. It is now reclassified as 'dwarf
planet'. It also has five satellites revolving round it.*

with respect to their composition. They form two distinct groups -
(i) the inner planets, and (ii) the outer planets.

Inner planets consist of Mercury, Venus, Earth and Mars. They are
also known as **Terrestrial planets** due to their unique composition and
rocky bodies. They are smaller in size. They are composed with large
proportion of heavier minerals like silicate, iron and magnesium. They
have greater specific gravity generally more than 5 except for Mars which
has a specific gravity of only 3.9. On the other hand, the outer planets are
much bigger in size e.g. the biggest of them; the Jupiter is about 11 times
bigger than our Earth. Even the smallest of them the Uranus is about 3.4
times bigger than the Earth. They, therefore, are also called **Giant** or
**Jovian** planets. They are mostly composed of light gaseous elements like
hydrogen, helium, methane and ammonia. Their specific gravity varies
between 0.69 for Saturn to a maximum of 1.71 gm /c$^3$ for Neptune.

The solar system also consists of a number of **asteroids** mostly
between the orbits of Mars and Jupiter. They are also known as **planetoids**.
Their composition resembles the inner planets and they like any other
planet also revolve round the Sun.

All of the planets except Mercury and Venus also have varying number
of **natural satellites** or moon under the field of their gravity e.g. earth

has one satellite, the Moon; Mars has 2; Jupiter- 67; Saturn -62; Uranus-27 and Neptune- 14. Numerous small satellites have been discovered in recent years during space explorations. There are almost 181 satellites of these planets and asteroids. They equally are the part of our solar system.

There are numerous **comets** and **meteoroids** which also move round the Sun and form part of our solar system. Comets are the bodies which are formed of extremely light materials such as stellar dust and gaseous materials. They consist of two parts- a very bright head known as **'coma'** and a broom like tail. Nucleus of the comet's head is composed of solid matters and frozen gases. They have a very **eccentric orbit** around the Sun. They do not have a very definite revolutionary period. Most of them have an orbital period of over ten thousand years. Very few of them like **Halley's Comet** have a revolutionary period ranging between 40 to 80 years.

Almost similar to comets there are **meteoroids** in our solar system. They are small rock or particles of debris in our solar system. They range in size from dust to around 10 metres in diameter. They move at a very high speed. Fastest ones may move at a speed of more than 42 kilometres per second. If they happen to enter the atmosphere of the Earth they are burnt due to friction and are vaporized. They are known as **meteors** or **shooting stars**. If parts of the meteoroids are broken and manage to reach the Earth's surface they are called **'meteorites'**. Analysis of the composition of meteorites helps in understanding the evolution of the universe. They also provide an insight in the internal structure of the terrestrial planets.

All the members of the Solar System revolve round their central bodies in an anti-clockwise direction in fixed times and on a definite path known as **orbit**. Planets, asteroids and comets move around the Sun and satellites around their parent planets. Orbital motion of these celestial bodies is the result of universal gravity as enunciated by Newton. The size of orbit of different members of the system varies (the causes of variation in motion of planets and size of orbits* is explained by Kepler). **Revolution period** of planets, therefore, is proportional to the size of their orbit a dependent variable to their distance from the Sun. Relative mass of neighbouring planets, however, disturbs the regularity of the planetary orbits. It causes oscillation about their **mean ellipse**. As a result sometimes they are relatively closer and sometimes farther from the Sun.

All planets also have another motion besides the revolution around the Sun. It is characterized by their movements on their axis. **Axis** of a planet refers to the imaginary line that passes through their North and South Poles. All planets rotate/spin on their axes at a uniform rate. Generally they complete their rotation in less than 25 hours (Earth time) with respect to the Sun. Some planets rotate very fast and some slow. Completion of one rotation on their axes broadly is termed as **sidereal day**[2]. Table-1 shows the period of revolution and rotation of different planets relative to Earth time.

Most of the planets and satellites rotate in the direction of revolution i.e. they move on their axis in counter clock wise direction except for Venus and Uranus. They are said to have **'direct** or **prograde rotation'.** The planets and satellites that rotate counter to the direction of revolution like Venus and Uranus and many of the natural satellites are said to have **'retrograde rotation'.**

**Table-1.1: Period of different motions of planets relative to earth time**

| Period | Revolution Period (in earth year) | Revolution Period (in earth days) | Rotation (Earth Time) |
|---|---|---|---|
| Mercury | 0.24 | 87.97 | 59 days |
| Venus | 0.62 | 224. 70 | 243 days |
| Earth | 1.0 | 365.26 | 23hrs 56 minutes |
| Mars | 1.88 | 686.98 | 24hrs 37 minutes |
| Jupiter | 11.86 | 4332.59 | 9 hrs 50 minutes |
| Saturn | 29.46 | 10,759. 22 | 10 hrs14 minutes |
| Uranus | 84.06 | 30,685.4 | 23 hrs 54 minutes |
| Neptune | 164.82 | 60189.0 | 18 hrs* <br> * (not very certain) |
| Pluto | 247.7 | 90,465.0 | 6 days 9 hrs |

## Chapter-1:2 (B) Measurements of Astronomical Distances

Interstellar as well as intergalactic distances are enormous. Measurement of these distances in terms of mile or kilometre or any measure like them will give almost incomprehensible numbers to remember or write.

Therefore, a different measure is adopted. Astronomical distances are very often measured in terms of **light years**. A light year is a measure denoting the distance travelled by light in one year. We know that the light travels at a speed of 298,000 kilometres per second. Therefore, a light year is equivalent to *298000 km.x60 seconds x60 minutes x 24hrsx 365 days i.e about 9.4 trillion kilometres.* **Proxima Centauri** the nearest star from our solar system is situated at a distance of about 4.2 light years. The nearest galaxy from our solar system, the *Canis Major Dwarf*, is located at a distance of about 25,000 light years from it.

Another measure adopted by scientists for astronomical distances is **parsec**. A parsec is equal to about 3.26 light years or 206 265 **astronomical units** (one astronomical unit being the average distance between the Earth and the Sun). If a parsec is converted into kilometres it will be equal to 3.086 (angle of one second arc subtended by one astronomical unit) X $10^{13}$ km. Shape, Size and Interplanetary Distances.

All the planets in our solar system though different in sizes are more or less spherical in shape. But they are not exactly spheroids except for

**Table-1.2: Distance of planets from the Sun and their physical dimensions**

| Planets | Distance (in million km) | Distance (in AU) | Diameter ( in km) | Mass (1023 kg) | Density (g/cm3-) |
|---|---|---|---|---|---|
| Mercury | 69.7 | 0.47 | 4,878 | 3.3 | 5.4 |
| Venus | 109 | 49.30 | 12,102 | 48.7 | 5.5 |
| Earth | 152.1 | 0.73 | 12,756 | 59.8 | 5.3 |
| Mars | 249 | 1.0 | 6,787 | 6.4 | 3.9 |
| Jupiter | 815.7 | 1.67 | 142,984 | 18,991 | 1.3 |
| Saturn | 1507 | 5.45 | 120,536 | 5,686 | 0.7 |
| Uranus | 3004 | 10.07 | 51,118 | 866 | 1.6 |
| Neptune | 4537 | 20.08 | 49,660 | 1030 | 1.2 |
| Pluto | 7375 | 30.33 | ? | ? | ? |

the Mercury and the Venus which are not flattened due to operation of centrifugal force under their rotational motions. Except for them, no other planet exactly resembles any known shape. They are like their own shape. The term used to describe such a phenomenon is **geoids.** In order to avoid complications in measurement of distances within inter solar system use of Astronomical Unit is made. An Astronomical Unit (AU) is the time taken by the rays of the sun to reach the Earth. It takes the rays of the Sun about 8.3 minutes to travel a mean distance of about 149.6 million km to reach the earth. The distance between the Sun and the Earth has been standardized to be 1(one) AU. Table-1:2 shows the comparison of the planets in different aspects and distance from the Sun including the now discarded planet the Pluto:

## Chapter-1:2(C): Composition of the Sun and the Planets

### *The Sun*

Sun along with its system of planets, satellites and other celestial bodies is believed to be about 4.6 billion years old. It is the only **self-luminous body** in our solar system and is the source of almost all the light and heat for the surfaces of all bodies in our planetary system. It is almost 110 times bigger than the radius of the Earth. It has a radius of about 700,000 km. On stellar heat scale the Sun is only moderately hot. In fact it is categorized under the 5th category of the hottest stars in the known universe and is only a dwarf in the families of stars. If one uses the **Morgan- Keenan system** of star classification our Sun is classified as **G2V star**[3]. However, the Sun alone accounts for about 99.87 percent of the total mass of the solar system that includes all the planets, satellites, asteroids, comets and meteors.

The sun like our planets has two motions. It rotates on its axis in about 27 days of earth time. However, rotation of the Sun is characterized by differential motion at different solar latitude. The solar equator takes about 26 days to make a full rotation whereas around 600 of solar latitude it takes about 31 days. This differential rotation may be attributed to the gaseous composition of the Sun. It causes formation of strong magnetic field around it. The Sun like planets also revolves but around the core of our galaxy- the Milky Way. It takes the Sun about 200 million years to complete its one revolution.

The Sun is composed mostly of gases. The gases that compose it are hydrogen and helium with only traces of heavier elements towards its centre. Therefore, despite its enormous size relative to Earth it has a mean density of only 1.4gm/cm3 against 5.5 of the Earth. Yet it is the most important source of energy to all the celestial bodies under the solar system. The Sun gets its energy from transformation of hydrogen nuclei to helium due to continuous **thermonuclear fusion**. Very little is known about the internal structure of the Sun. However, some insight may be had from the study of phenomena associated with its upper layers.

Composition of its upper layer is characterized by superposed but highly active heterogeneous layers. The upper most layer is **corona**, a halo of white light visible only during the solar eclipse. Inner coronal area extends about to an area equivalent to two solar radii beyond the solar disc. Corona also forms the part of the very little understood **solar winds**. It extends up to 3 million km in **inter galactic interplanetary space** as the part of solar wind. Solar wind blowing at supersonic speed covers whole of the interplanetary space till it mingles in the intergalactic space beyond.

The second layer of about 2000 km is **Chromosphere**. It is visible again only during the solar eclipse as a colourful ring surrounding the black disc that covers the Sun. Otherwise both of them may be seen on a normal day only by the use of specific spectrum technology using Halpha and X-rays respectively. The visible layer of the Sun, in normal circumstances, is **Photosphere**. Almost all the solar energy in the form of **electromagnetic energy** is emitted from this layer. The upper layers of chromosphere and corona though are intrinsically linked with photosphere, exhibit uniquely distinct temperature conditions when compared to other planets of the solar system.

Temperature in chromosphere suddenly increases from 4300 k at the top of the photosphere to more than10, 000 k. In coronal disc around the chromospheres temperature further rises to more than 1 million k. It may be noted that the temperature in other planets normally decrease from lower layers to the outer layers.

The surface of the Sun-the photosphere, is composed of white grains which move constantly due to convective motion of matters from its

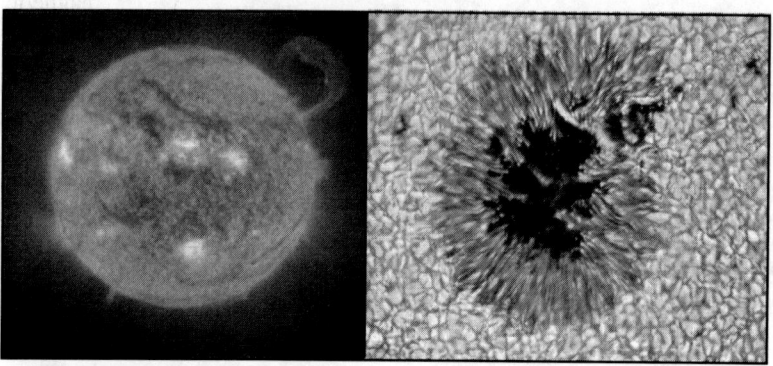

(A) Distant view of solar surface    (B) Closer view of the solar surface

**Figure 1:5: Solar surface exhibiting granular composition**

central core (Fig. 3). Granular composition of the layer is attributed to the outward motion in the underlying convective zone. These grains 1000-2000 km in size have a life of about 10 minutes and are continuously consumed by downward moving section of Figure 5: Solar surface exhibiting granular composition the convective currents.

Knowledge about the interior composition of the Sun is only hypothetical. It is estimated that about 600 million tons of hydrogen nuclei is converted to helium every second nearer the core of the Sun. It is also believed from the observation of outer layers that temperature and specific gravity of the central core may be as high as 15 million degrees Kelvin and 160 gm/cm$^3$ respectively. Temperature and density from the centre outwards decreases very rapidly. Number of hydrogen nuclei correspondingly increases rapidly outwards. Effective temperature of outer most normally visible layer- photosphere is believed to be 5770 k.

Temperature even in photosphere is not uniform. It may have variations in its surface temperature to the magnitude of about 1500°C 2000°C. The cooler regions are seen as spots on the solar surface. They are known as **'Sunspots'**. Sunspots generally have two componentsdarker but cooler centre and lighter but relatively warmer peripheral areas. The darker centre is called **'umbra'**. The lighter sector is called **'penumbra'**. Sunspots vary in size and may occasionally have a diameter many times bigger than that of our Earth. They occur probably due to temporary absence of convective heat in the region. They may exist for many solar rotations. The darker centre of the Sunspot, however, is stirred by movements of matter by upcoming convective currents from the solar core. They are associated also with very bright patches and big explosions. The bright patches are known as **'faculae'**. The big solar explosions are known as **'solar flares'**. Numbers of sunspots and associated phenomena of magnetic field reversal vary between nine to twelve and half years. The sun spots numbers from one maximum to the other, thus, takes about 22 years. When the incidence of sun spots takes place more profusely it is known as **'solar maximum'**. One of the sunspot maxima ended only in 2012. The present cycle is on decline. Next maxima may be expected around the year 2023 and 2044 in a sunspot cycles of 11 and 22 years respectively. Inversely, when occurrence of sunspot and associated solar eruptions are least it is known as **'Maunder minimum'**. And though there is no direct evidence, maunder minimum is recorded to have coincided with the cold spells on the Earth. More recently it happened between 1650 and 1700 AD.

Sun spots are also intrinsically related with the **solar magnetic field** which controls its inner convective currents and, thus, the incidences of sun spots. Strength of the magnetic field also impinges on the occurrence of **aurarae borealis** and **aurarue australis** near north and south poles of the Earth. Solar surface is also characterized by the incidences of explosion caused by continuous 'thermonuclear reactions' in the lower chromospheres. The phenomenon is believed to cause gigantic flares and deformation in emission of ultraviolet rays and changes in electromagnetic field. The impact of these eruptions is also visible as tongue like flames in coronal sphere. It is caused by embedding of high density gases in low density lower corona.

This is known as **'prominence'**. These flares travel at a speed of 1000 km per second and may reach up to 7,000,000 km (10 solar radii) i.e engulfing almost the entire solar system. Though not fully understood, by implication it may increase the temperature ushering in a warmer phase on the planets including the Earth. Recent warming of the Earth atmosphere may at least partly be attributed to this phenomenon and associated bombardment of solar radioactive and plasmatic matters on the Earth. They are also believed to disturb the ionosphere of the Earth hence radio waves and telecommunication systems operating in our world.

## Chapter-1:2 (C): Composition of The Planets

As stated above (see chapter-1:1(A)) the two groups of planets distinctly vary in their composition and accordant dispensations. Though formed from a common source they mineralized at different distances from the Sun hence at different temperatures. The inner or the terrestrial planets comprising the Mercury, the Venus, the Earth and the Mars along with the asteroids show solid and rocky bodies. The Earth is the biggest amongst them with a radius of about 6371 kilometres followed in order by the Venus, the Mars and the Mercury (see table-1:3). They are composed of relatively heavier elements consisting mainly of silicates, iron and magnesium. These components are distributed in an uneven manner from surface towards their central core. They reflect different layering with variable thickness. The three planets nearest to the Sun (Mercury, Venus and the Earth) have larger concentration of heavier elements. Therefore, their density of above 5.5 grams per cubic centimetre ($gm/cm^3$) is also the highest in the solar system. Mars, the fourth from the Sun has relatively lesser presence of heavy elements. Therefore its specific gravity at 3.9 $gm/c^3$ though higher than the outer planets it is distinctively low as compared to other inner planets. The Moon, the only satellite of the Earth also shows characteristics of terrestrial planets though with a little lesser concentration of heavy elements in its body. The density of Moon is about 3.34 gm/c³. It is a little lower than that of Mars.

Despite various attempts to understand the interior composition of these terrestrial planets information are very sketchy and inconclusive. However, there is general agreement in the community of geophysicists that all these planets have layered structures like that of the Earth but of different dimensions. Table-1:3 gives an idea about the thickness of different layers of terrestrial planets modelled after the Earth.

**Table-1.3: Showing densities, radiuses and estimated thickness of different layers of inner planets**

| Terrestrial Planets | Density | Radius (In km.) | Crust (In km.) | Mantle (In km.) | Core (In km.) |
|---|---|---|---|---|---|
| Mercury | 5.42 | 2440 | ? | ? | ? |
| Venus | 5.25 | 6050 | ? | ? | ? |
| Earth | 5.52 | 6371 | 35 | 2900 | 3471 |
| Mars | 3.94 | 3395 | 50 | 2045-1345 | 1300-2000 |

The outer giant and gaseous planets - Jupiter, Saturn, Uranus and Neptune (Pluto may also be included in this category despite a different origin) are primarily composed of very light gaseous elements. Predominance of these elements is found to have been effected by their distances from the Sun. The Jupiter and the Saturn which are relatively closer to the Sun amongst the outer planets have a predominance of hydrogen and helium in their composition as in case of the Sun. On the other hand, Neptune and the Uranus which are a little farther from the Sun are mostly composed of methane, ammonia and water ice. As such their specific gravity is much lesser than the inner planets despite their sizes (see table-1:2 and compare their size and densities).

All planets other than the Mercury possess some kind of atmosphere. However, there is no clear cut division of structure in case of outer gaseous planets. But in case of inner planets it is believed that in the process of their consolidation some gases in the form of water vapour ($H_2O$), carbon dioxide ($CO_2$), carbon monoxide (CO), and nitrogen ($N_2$) were released due to degassing of solid surface materials. Hydrogen and oxygen in the atmosphere were obtained through decomposition of water when it was hot. However, initial atmosphere of inner planets did not have free oxygen as it was almost immediately synthesized in their terrestrial rocks.

The mass of the planets, their specific gravity and distance from the Sun are the important factors to affect their composition and formation of their atmosphere. For example, the Mercury, nearest to the Sun having a very low mass, thus a very weak gravitational field, does not have an atmosphere. It is because of the fact that the escape velocity (a reflection of the strength of gravity) of matters on this planet is only 4.3 kilometres per second (km/second) against 11.2 km/ second on the Earth. A very high surface temperature due to its nearness to the Sun makes the degassed components from its territory highly sparse and mobile to escape its field of gravity. All other planets having a stronger gravitational field have been able initially to arrest constituents of their present atmosphere from primordial stellar dusts and gases.

This explains the variable nature of surface configuration as well as inhomogeneous layering of different inner planets. The process of degassing in the wake of variable nature of planetary consolidation (solidification in case of inner planets) explains the variable composition of atmosphere around different planets. It needs to be underlined that the chemical composition of gases released into the atmosphere depends mostly on two factors: (i) temperature and (ii) state of oxidation of the rocks. Thus, the atmosphere around the Venus and Mars has an abundance of Carbon dioxide ($CO^2$) in gas form whereas in case of the Earth there is an abundance of nitrogen and oxygen in its atmosphere wherein most of $CO^2$ has been fixed in the form of calcium carbonate in its water and rocks. Why the atmosphere of the three closest terrestrial planets is so different is still not fully comprehended. Yet, it may be said that the atmosphere of these terrestrial planets have followed different evolutionary paths. It is also reflected in the receipt of solar energy particularly in the form of ultra violet radiation. It is the presence of nitrogen, carbon dioxide and little hydrogen, water vapour and carbon monoxide in combination with ultra violet radiation and lightening which is believed to have initiated the evolution of life forms. It is these considerations that propel scientists to search for life forms- present or past on Venus or Mars. Presently, however, it is only the Earth which has been capable of supporting life forms for last 3.7 billion years.

## Review Questions

1.  Define Universe. What are the basic elements from which the Universe is believed to have emerged?
2.  What are the major school of thoughts with regard to the origin of the Universe? Which theory about the origin of the Universe is more convincing to you and why?
3.  Assess the contributions of Copernicus, Kepler. Galelio and Hubble in explaining the mysteries about the Universe.
4.  Describe the units of measurement generally used to measure cosmic distances.
5.  What is a solar system? What bodies constitute our solar system?
6.  How is the energy emitted from the sun produced?
7.  Describe the structure and composition of the Sun. Fundamentals of Physical Geography.
8.  What is Morgan-Keenan system of star classification?
9.  What do you understand by terrestrial and Jovian planets? Describe the composition of different planets.
10. Why Pluto is not considered to be a member of our solar system?
11. Name the members which make solar family.
12. Explain the following:-

    (a) Sun spot; (b) Solar flare; (c) Umbra; (d) Corona; (e) Chromosphere; (f) Photosphere; (g) Rotational motions of the Sun; (h) aurarae borealis and aurarue australis; (i) Maunder minimum; (j) Types of galaxy; (k) Geoid and spheroid; (l) meteors and meteorites (m) solar maximum.

## Reference

1.  Audouze, J. and Israel, G (Eds) (1986): The Cambridge Atlas of Astronomy, Cambridge University, Cambridge.
2.  Hubble E, (1925): cited in Audouze, J. and Israel, G (Eds) (1986): The Cambridge Atlas of Astronomy, Cambridge University, Cambridge; p. 332.
3.  Christian, D. (2011): Maps of time: an introduction to big history; University of California Press, Berkeley; p. 671.
4.  Hawking, S. (2011): The Theory of Everything: The Origin and Fate of the Universe; Jaico Pub.
5.  House; Ahmedabad; p. 132.

## Notes

1.  Pluto considered being the ninth planet till 2006 is not considered a solar planet under the new definition of planets by International Astronomical Union. Instead its solid icy body is considered to have been captured by the gravitational pull of our Sun from a disc shaped region of the Kuiper belt that exists in the neighbourhood of Neptune. Unlike rest of the outer planets Pluto is almost the size of our earth and cannot be considered under the group of giant planets.
2.  sidereal day refers to time taken by planets to complete one rotation.

3. The Morgan–Keenan (MK) system uses the letters O, B, A, F, G, K, and M, a sequence from the hottest (O type) to the coolest (M type). Each letter class is then subdivided using a numeric digit with 0 being hottest and 9 being coolest (e.g. A8, A9, F0, F1 form a sequence from hotter to cooler. Roman numerical from 0 to IX denotes the relative size of stars from hyper giants to brown dwarfs. Name Revolution.

# Chapter - 2

# Solar System

## Objectives

Main objective of this chapter is to make students aware of the developments that have taken place since ancient times with regard to the origin of our solar system. It is devoted to bring about clarity to its users aware about the changes in approaches to the study particularly after 15th century. A comprehension of obtained information is expected to enable the users to correlate the ever changing processes and their resultants on our planet.

The chapter undertakes the study in respect of:

- The established view points of ancient scholars in relation to the scientific revolution of 15th century.

- General characteristics of our solar system

- Over view of different proposed hypotheses with regard to suggested mechanism and their ability to explain different characteristics of our solar system.

## Chapter 2:1 - Origin of the Solar System

Origin of the solar system has been a mystery since time immemorial. Most of the early scholars left the subject under the domain of theology. Orthodox theologians who believed in 'creation' attributed it to God or a super natural power. They also believed in a **geocentric solar system** with earth at its centre. Scholars like **Aristotle** (384-322 BC) and **Ptolemy** (2nd AD) also supported this view. The view despite objections from many ancient and medieval scholars was accepted by large number of people for over 1500 years. In recent times, following the findings of **Copernicus** (1473-1543) and **Kepler** (1571-1630), **Descartes** (1632) and **Swedenborg** (1734) reintroduced the idea of evolution and attempted to provide a scientific explanation of the origin of the solar system. Descartes, considered to be the father of modern **cosmogony**, believed that the Sun and planets had condensed from masses of swirling particles with which the Universe was initially filled. The particles, in course of time contracted and condensed to form the Sun and the planets having a circular motion. As **law of gravity** and **law of motion** were not known to him (they were propounded by **Newton** (1642-1727) he was unaware that matter did not behave in the fashion he visualized. Swedenborg in 1734 liked to believe that all the planets essentially were ejected from the Sun during particular violent period of sun-spot activity and were closely associated with **solar prominences**. **Buffon** (1749) attributed the evolution of the solar system to a collision of a comet with the Sun causing ejection of a solar filament and which might have condensed in a system of present planets. Better understanding of the nature of comets in the following years rejected his hypothesis. But it successfully laid the foundation of strong catastrophic and dualistic viewpoints in cosmogony. Not satisfied with the dualistic explanation **Immanuel Kant** in tune with Descartes' view proposed in 1755 his **monistic theory** of evolution of the solar system from a gaseous nebula. Explanation about the origin of the solar system in general and the Earth in particular have been given from several viewpoints since. Many authors, however, describe the most recent ones or the most widely known hypotheses, to the exclusion of all others. It, however, needs to be realized that no theory so far has been able to explain all the attributes of the solar system completely. Increasingly improved technology and better scientific understanding of planetary and space phenomena and their logical extension in the field of information has made it almost mandatory to revisit earlier postulates on which contemporary theories have been constructed or are being constructed. Recent studies, in this light though may be said to favour nebular hypotheses of yester years, explanations obtained from catastrophic hypotheses have not been completely disregarded. In fact, no view point so far has been found to explain the origin of the solar system in its entirety. In order to arrive at a satisfactory explanation in respect of the origin and evolution of the solar system, it is felt that it must address the known general characteristics of the solar system as listed below:

(i)  Different members of the solar system (Sun, Planets. Satellites, asteroids, comets etc.) were formed almost simultaneously some 4.6 billion years ago.

(ii) Entire physical properties of the solar system were obtained from the same source.

(iii) Each member of the solar family despite great distances has their own independent and almost circular orbital path around the Sun.

(iv) All planets within the solar system are confined to same disk i.e. the plane of ecliptic, close to the equatorial plane of the Sun.

(v) There is an abundance of elements like deuterium and lithium in all planets but not in the Sun.

(vi) Density and chemical composition in the planetary system vary. The inner terrestrial planets are composed mostly of denser silicate, magnesium and iron. On the other hand, the outer giant gaseous planets are composed of different light cold gaseous materials and frozen ice e.g. there is an abundance of hydrogen and helium in case of Jupiter and Saturn but in case of Uranus and Neptune methane and ammonia gases dominate.

(vii) Revolution and rotational speed of planets is faster than the Sun.

(viii) There exists a sequential and orderly arrangement of planets with regard to their size and number of satellites.

(ix) Most of the planets have direct rotation i.e. they rotate along the direction of their revolution which is anti clock wise except for the Venus and the Uranus which have retrograde rotation in clock wise direction i.e opposite to the direction of their revolution.

(x) Most of the planets have a sidereal period of the rotation less than 25 earth hours (Jupiter and Saturn about 10 hours, Uranus and Neptune about 15 hours, Earth and Mars about 24 hours) but nearest of the planets to the Sun the Mercury and the Venus has a sidereal period of 59 and 243 earth days respectively.

(xi) Greater part of the mass of the solar system i.e. 99.87 percent is concentrated in the Sun with only a mass of 0.13 percent in all the planets put together.

(xii) The Sun conversely has a very low angular momentum of about 2 percent of the total planetary system. More than 98 percent of the angular momentum of the system is accounted for by all the planetary bodies. (Point xi and xii goes against the presumption that all the planets were born out of the present Sun representing the primordial nebulae)

## Chapter - 2:2

Any attempt at theorizing the origin of the solar system as well as of the Earth, therefore, must attempt to provide a valid explanation of these characteristics of the system. And though many views have been expressed since 2nd half of the 17th century no single explanation has been able to satisfy all the queries in respect of the unique phenomena that the solar system is. Partially, however, the scholars have been able to explain quite a few phenomena based on their observations and experimentations in space which are generally obtained from the understanding of physical law as understood to be operating on the Earth and within the solar system.

The Views regarding the origin (evolution) of the solar system may be said to have two major approaches- (a) **Monistic** and (b) **Dualistic**. Monistic approach as the name suggests primarily is based on the premise that the solar system had a parent and evolved from a single star or a nebula. This is also known as **parent hypothesis**. There have been many protagonists of this class of hypotheses since 1755. They include scholars like **Immanuel Kant** (1755), **Pierre-Simon Laplace** (1796), **Lockyer** (1890), **Alfred Hoyle and R.A Lyttleeton** (1939), **Hannes Alfaven** (1942), **Von Weizsacker** (1943), **Otto Schimdt** (1943), **Gerald Kuiper** (1951), and **Safranov** (1971).

**Dualistic** or **bi-parental** approach attributes solar system to interaction between two celestial bodies which might have come closer to set a chain of events culminating in the formation of the members of the system. The proponents of this view points since 1745 include persons like **George Buffon** (1749), **Chamberlin and Moulton** (1904), **James Jeans** (1910) and **Harold Jafferys** (1926), **H.N. Russell** (1937) and **Lyttleton** (1938), **Ross Gun**, and **A.C. Bannerjee** (1942).

One may also classify the views about the evolution of our solar system broadly in three categories as follows:

**Catastrophic theories**: They include Dualistic, Big Bang and Nova hypotheses as well as planetesimal hypothesis of Chamberlin and Moulton. They may be grouped as having two premises. One group of these theories are constructed upon the assumption that two celestial bodies one of which must be the Sun might have come closer and set a sequence of events leading ultimately to the formation of planets and other members of the solar system. Such events as the existing knowledge of galactic space suggests though are very rare are definitely not unknown.

The other group of hypotheses under this group are built upon a belief that the solar system as a whole originated from catastrophic incidence during the evolution of a nebulae and mechanism of which is closely related with the evolution of stars. Big Bang theory as proposed recently is one such view point.

**Normal stellar evolution Theories** including gaseous and nebular hypotheses, and Theories of unique stellar evolution that include theories of electromagnetism of Hannes Alfaven, Inter stellar Dust of Otto Schmidt, Nebular Cloud of Von Weizsacker, Proto planet of Gerald Kuiper and collision- Accretion hypothesis of Safronov.

All groups of hypotheses are able to justify some characteristics of the solar system but fail to explain many others. It is, however, essential to understand the basic premises and their explanations to comprehend the processes and their resultants which might have led to the formation of our solar system. In following section an attempt has been made to sequentially explain some major hypotheses under different categories.

## Chapter - 2:2:1 The Nebular Hypothesis of Kant and Laplace

One of the first attempts to explain the origin of the solar system in modern times was made by monist philosopher and German scholar Immanuel Kant around 1755. He laid the foundation of his hypothesis on the Newtonian law of gravitation. He, in the process, also suggested as to how a nebula was formed from primordial gaseous cloud. His view about the origin of the solar system, thus, is also known as *gaseous hypothesis*. His presumptions and suggested mechanism of the formation of the solar system may briefly be outlined as follows:

- Initially the space was filled with motionless clouds of primordial particles.
- These particles under mutual pull of gravitation started to acquire directionless motion.
- They, thus, started to collide and accrete with each other leading to fragmentation of interstellar clouds.
- Collision and accretion of particles generated more heat and rotational motion.
- In course of time the process converted the erstwhile motionless cloud of particles into a hot rotating nebula.
- Rotating nebula started to contract under the influence of its own gravity and loss of heat from its outer sphere.
- Because of viscosity and increasing centrifugal force the contraction of nebula led to the formation of disk along its equator in the form of rings.

- Rings of the cooler outer disk subsequently contracted to form planets.

- Remaining part of the hot central core of the nebula still rotating remained as our Sun.

- Still under the gravitational pull of the central mass, the Sun, these planets started to revolve round it.

- The same process was invoked by Kant to explain the formation of satellites in which initially hot rings were separated from the planets. After being consolidated these satellites started to revolve due to gravitational pull round their respective parent planets.

Laplace, a French mathematician put forward his nebular hypothesis of planetary evolution in1796 though similar but independent of any knowledge of Kant's view on the subject. He unlike his predecessor was influenced by the knowledge of (a) existence of a nebula in the universe, and (b) sighting of rings around the planet Saturn by Galileo in 1610. In his explanation of the formation of the solar system he, unlike Kant, assumed the presence of a massive hot and already rotating nebula of rarefied matter. He then outlined the sequential changes and events that might have influenced the formation of the solar system as follows:

- The size of the massive primordial nebula started decreasing in its size due to contraction caused by relatively higher rate of cooling of the outer sphere.

- Decrease in size of the nebula would lead to faster rotational motion in it (this is believed to be in accordance with the law of motion) giving the nebula a flattened, disk like shape.

- With greater velocity the centrifugal force on outer sphere would start increasing till zero gravity is attained at the equator of the nebula. At the same time outer sphere would be losing its heat by radiation.

- Further increase in centrifugal force would make the outer sphere of the nebula on its equator weightless.

- Cooler outer sphere could not remain integrated with still cooling inner nebula due to setting in of compositional changes.

- In time, rings of gaseous matter became separated from the outer part of the nebular disk, until the smaller nebula at the centre was surrounded by a series of rings (till then only seven planets were known so Laplace talked only about seven Saturn like rings around the nebula).

- Out of the material of each ring a great ball was formed, which by shrinking eventually became a planet.
- In the same fashion satellites are believed to have been formed around their parent planets. This also explains why larger planets have larger number of satellites.
- Remaining central part of the nebula continues to be our Sun.

Evolution of the solar system according to the nebular hypothesis may well be understood by the following representative diagram:

*Source:* utk.edu/astr161/lect/solarsys/nebular.html

**Fig. 2.1: Evolution of Nebulae.**

## Merits of the nebular hypothesis

This hypothesis is found to be in tune with quite a number of the observed facts about the Solar System. It is able to broadly explain-

- The existence of planetary orbital paths more or less in a plane coinciding with solar equatorial plane.
- Revolution of planets in the same direction.
- Rotational motion of most of the planets in the same direction.
- Perpendicular rotation axes of nearly all the planets to the orbital plane.
- One of the greatest merits of the hypothesis lies in the fact that it has again invoked the nebular hypothesis in one way or the other.

## Critcism

Despite its popularity for more than 100 years till early parts of the 20th century the hypothesis is, however, found untenable in the light of recent calculations and findings. They may be listed as below:

- Kant's assumption of cold and static particles colliding with each other under mutual gravity and generating heat and rotation is considered to be against the **law of** conservation of angular momentum.

- His assumption that there was an increase in the velocity of rotation with the increase in the size of accreting primordial matter is also against the above law. In fact speed of rotation is inversely proportional to the size of the rotating mass of bodies.

- Distribution of angular momentum in the planets and the Sun does not match the distribution of mass and distance.

- Condensation of ejected rings from the central disk of the nebula into planets is also untenable because the gaseous rings would dissipate before they are condensed in planetary form.

- The mass of the ring formed around the nebular central mass according to the British. Physicist **J.C. Maxwell** would be so small that they could not condense into planets under the gravitational attraction.

It must, however, be realized that many information about the planetary bodies and the space which are available today were not available during the time of Kant and Laplace. They, still, may be accredited with first serious attempt to explain the origin of the solar system invoking an established law of gravity. They may also be given credit to be the forerunner of many subsequent hypotheses.

## Chapter - 2:2:2 - The Planetesimal Hypothesis

Planetismal hypothesis as proposed by Chamberlin and Moulton in 1904 provided an alternative mechanism to explain the formation of planets and their satellites to address the criticism against the nebular hypothesis. The mechanism of planetary formation proposed by them is also considered by many to be a serious attempt to explain the internal composition of terrestrial planets and the earth's atmosphere.

Their hypothesis appears to have been inspired by Buffon's view of comet-sun collision suggested in 1745. It likewise is built on the premise that the Sun did not have satellites initially.

It is the subsequent chain of events and encounter with other star in the interstellar space that created all the planets and other celestial bodies of the solar system. They outline the events as follows:

- The Sun which even today exhibits great explosive force encountered a massive but smaller star passing very close by. It enhanced solar activities.

- Under the gravitational pull of the star two simultaneous activities affected both of them-

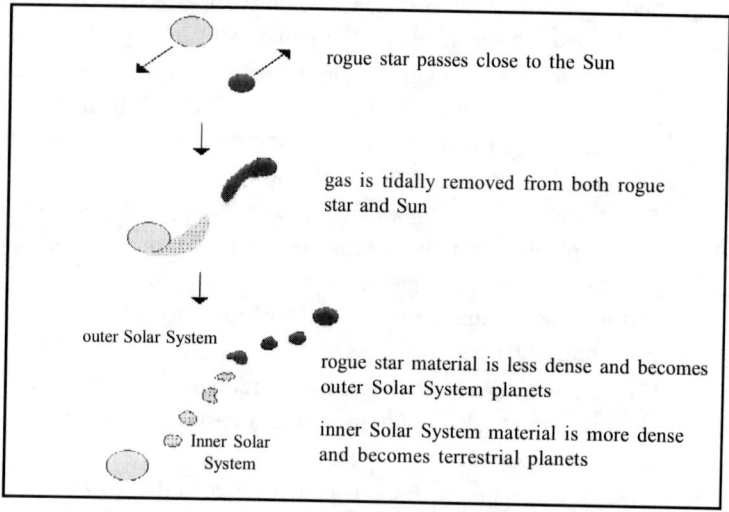

rogue star passes close to the Sun

gas is tidally removed from both rogue star and Sun

outer Solar System

rogue star material is less dense and becomes outer Solar System planets

Inner Solar System

inner Solar System material is more dense and becomes terrestrial planets

*Fig. 2:2: Formation of Planetesmals.*

- Solar matters were ejected with the prominences extending up to hundreds of thousand kilometres in the space reinforced by Sun's own explosive nature and

- Huge tides were created on the surface of both under mutual gravitational influence of each other. Solar explosions aligned with the intruding star were suggested to be more on its side. Having gravity differentiations large masses of matter were shot beyond the matters produced by the tides.

- Under the cross-pull from the rapidly fleeing star some parts of the tidal and exploded matters from the Sun remained under the gravitational pull of the Sun and were thrown into an elliptical orbit around it.

- The total mass of matter ejected from the solar surface was much more than the total mass of the present planets.

- It was assumed that a large part of such matters fell back on the surface of the Sun as the gravitational pull of the intruding star moving in interstellar space decreased with its increasing distance from the Sun.

- It is suggested that 10 such ejection of matters must have taken place when the intruding star was passing by.

- The smaller masses of material quickly cooled to become solid bodies having a radius of one kilometre to ten kilometres. They are named planetesimals by the pioneers of the hypothesis.

- These solidified bodies would have their own nucleus and gravity proportional to their sizes. It would lead to their movements on irregular paths causing collision of planetismals in their elliptical orbit around the Sun.

- Five large masses of light matters away from the Sun and nearer to the passing star aggregated to form the nucleus of outer planets. Further aggregation of matter of similar composition and their consolidation under their own gravity later formed the giant outer planets.

- Five smaller masses of denser matter remained closer to the Sun to form terrestrial proto planets with denser core and relatively less dense exterior.

- Further aggregation of small planetismals added to the growth of both inner and outer planets.

- Aggregation of left over small planetisimal under the gravitational influence of planets led to the formation of satellites.

## Merits of the hypothesis

Basic assumptions that planetesimals formed out of gaseous mass and provided nucleus for further accretion to form planets continue to be an important concept. The concept finds proper recognition in later hypotheses in which planets are suggested to have formed due to collision and accretion of planetesimals or planetoids.

It provides explanation for direction of the planetary rotation and revolution.

- It accounts for the eccentricities of planetary orbits.

- It provides an explanation for higher angular momentum in planets than the Sun.

- It accounts for the presence of comets and meteorites in the solar system.

- It attempts to provide an explanation for the composition of the planets both inner and outer.

This was probably the first serious attempt to provide rational explanation for the formation of atmosphere and oceans on the Earth.

## Criticism

- The assumption that a passing by star came closer to the Sun at a very high speed inducing separation of matters from solar surface as well as its own is believed to be untenable because the intruding star should not be farther than the diameters of the two stars to affect ejection of matter from the Sun. This is believed by the cosmologists to be such a rare event that no scientific explanation can be developed on this assumption.

- **Jeans** in 1939 opined that only a close approach of a star was sufficient to cause ejection of matter from the solar surface. There was no need to invoke solar prominences to effect ejection of solar matters as suggested in the hypothesis.

- To effect the ejection of matters from the solar surface the intruding star must pass by the Sun at a velocity of more than 4500 km. per second. Such a speed is considered to be beyond the **escape velocity** of the galaxy. It, therefore, is untenable that a star could come nearer to the Sun. In 1940s, **Russell**, an American astronomer himself, proved that the force required to produce the angular momentum of the Jupiter would throw the matters entirely out of the solar system.

- American physicist **Spitzer** believes that the ejected material from the sun instead of condensing to form the planets would rather dissipate.

- If the planets were formed due to accretion of planetesimals they would not have such a high angular momentum as they are having presently.

- Though nucleus of the proto planets were responsible for attracting planetismals, according to **Jaffery**, it was impossible for them to attract atmospheric elements.

- Even if it is assumed that there might have been some gases and water content in the nucleus of proto planets, accretion of planetismals would not allow them to come out to form atmosphere.

Though developed in opposition to the nebular hypothesis of kant and Laplace and discarded by the 1940s significance of the planetismal hypothesis lie in the fact that many understandings provided by it have been retained in many of the later postulates of the recently revived nebular hypotheses about the origin of the solar system.

## Chapter - 2:2:3 Gaseous-Tidal Hypothesis

James Jeans was critical of Planetismal hypothesis of Chamberlin and Moulton as well as of the earlier views of late 19th century which advocated that planetary matters having different composition were not derived from the Sun. Rather they were captured by the core of a spiral nebula of meteoritic dust from outer space. The Asteroids in between the orbits of Mars and Jupiter were considered under the capture theory as proposed by Sea in 1896 and further developed in 1911. They were considered by the protagonists of the capture theories to be the remnants of small planets with which the whole system was originally filled. Not agreeing with these propositions and particularly with the invocation of violent solar explosions to provide matters for the condensation of planetismals Jeans in 1919 proposed the gaseous- tidal hypothesis which was later modified by Harold Jaffrey in 1929. He also believed that the planets in our solar system though different in sizes are more or less spherical in shape. But they are not exactly spheroids except for the Mercury and the Venus.

The hypothesis of Jeans and Jeffrey suggests that the formation of the planets were caused by disruption of the primitive sun by the tidal effects of a much bigger and massive star passing close by it. They assumed that the intruding star was moving relatively slowly than what was proposed under planetismal hypothesis. They pictured production of a solar tidal bulge on the side nearest to the passing star as a long gaseous filament. It was believed to have been shaped like a Cigar having a bulge in the centre and tapering towards the intruding star and also towards the Sun. The hypothesis seems to have been much influenced by the **axioms** (self proven facts) about the solar system which includes (i) arrangements, mass as well as shape and size of the planetary bodies in our solar system, (ii) terrestrial and gaseous nature of the inner and outer planets, and (iii) distribution of the number of satellites. The course of events and their resultants as postulated under the hypothesis may be sequentially outlined as follows:

- A massive star much bigger in size passed by the Sun.

- Solar matters were disrupted under the tidal effects generated by the passing star. It produced a long extremely hot gaseous filament of solar matters on the surface of the Sun.

- Moving over the solar surface the filament increased in size with closer approach of the star to the sun.

- As the star journeying in inter- stellar space crossed the Sun the tidal filament was detached under the increased gravitational pull of the star. It extended beyond the orbit of the remotest planet.

- The filament formed due to tidal effect was pointed at both ends due to the influence of the gravity of the star at one end and the Sun on the other.

- When the visiting star faded away in interstellar space, its tidal pull ceased and no more of the sun's matter could be lost.

- The filament as a whole then came under the gravitational pull of the Sun.

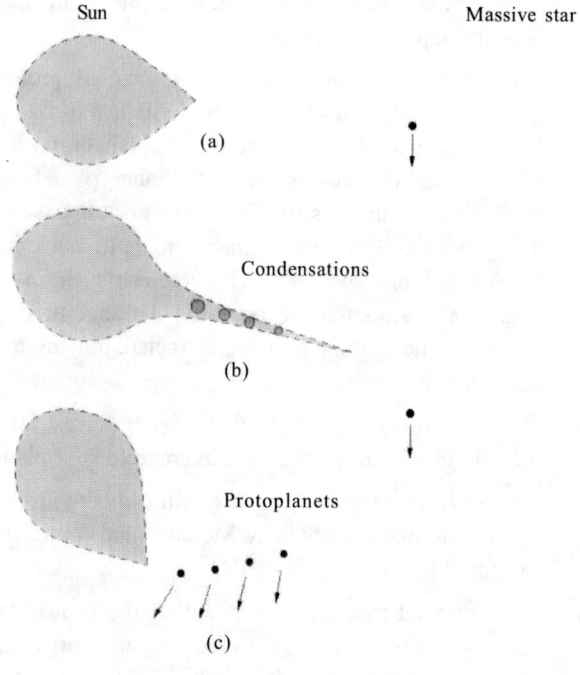

After David Darling (1999)

***Fig. 2.3: Gaseous- Tidal Hypothesis and***
***Evolution of Solar System***

- In course of time the matters within the filament started cooling.

- While cooling it condensed into sectors of planetary beads of the size of present planets.

- Condensed beads of planets started moving in their respective orbits around the Sun in the same direction in which the unbroken tidal filament was moving.

- One of the planetary beads is assumed to have been disrupted in the course of cooling to form the planetoids or asteroids in between present mars and Jupiter.

Following diagrammatic representation may illustrate the assumptions and their resultants as conceived by Jeans and Jaffery.

## Merits of the hypothesis

■ The hypothesis provides a reasonable account of distribution of size of the planets i.e. the larger planets would occupy the central bulge of the ejected gaseous filament and the smaller ones the tapering margins.

■ Internal gravity proportional to the size of proto planets thus formed would restrict dissipation of primordial gas particles. This may explain why the outer planets with larger mass continue to remain gaseous. The inner planets, on the other hand, might have suffered wastage of gaseous materials firstly because they were unable to hold back it due to their lesser mass, and secondly being nearer the Sun its great gravity might have attracted the escaping matters. In case of Mars it might be the Jupiter to have attracted the dissipating matters from it.

■ This is able to explain why the inner terrestrial planets have specific gravity more than the average for the planetary system.

■ It also explains the number of satellites being proportional to the size of the planets and why Mercury and Venus are not having any satellite.

■ It is believed that the movement of the proto planets through gaseous medium of the ejected filament providing resistance would retard the outward movement of the proto planets. This is expected to check the eccentricity of their orbits and make them more circular as it is found.

## Criticism

- Critics of the hypothesis suggest that the star that encountered the Sun was not only very big and massive but it must have come very close to the Sun to cause ejection of the gaseous filament. It defies the collected astronomical data.

- Greater difficulty lies with the explanation of rotational motion of the Sun which under the hypothesis is attributed to falling back of the parts of erupted tidal and filament matters on the surface of the Sun and the planets respectively particularly in case of Jupiter. Jaffery himself found it difficult to explain it. He gave more credence to collision theories to explain the rotations as explained under planetismal hypothesis.

- Hot gaseous matters in the filament having high momentum are more likely to dissipate and lost in the space instead of being condensed into planets.

- If planets are formed of solar tidal matters then why planetary composition differs so much from that of the Sun.

- Scientists like **Parisky** find it difficult to explain the distances of the planets from the Sun under the basic postulates of the hypothesis. **Russell**, the American astronomer, also suggests that according to the hypothesis planets should have been much closer.

## Chapter- 2:2:4: Binary Star Hypothesis

Despite various theories that were propounded to explain the origin of the solar system, no one could solve the problem of the great distances to which the planets have been formed. Many scientists though believed in basic tenets of the gaseous tidal hypothesis they attempted to overcome the criticism through the introduction of binary star hypothesis. This requires to be noted that the existence of binary system of stars was established by **William Herschel** way back in 1802(*binary stars denote to a system of two stars which revolve around a common centre of mass and are gravitationally bound*). Now it is established that almost 30% of the stars in the Universe are parts of a system of binaries. Prof. **H.N. Russell** in 1937 and **Lyttleton** in 1938 developed a hypothesis basing their assumption on Sun being the partners in one of such binary system. In the process they attempted to resolve the problem that discredited its predecessor- gaseous tidal hypothesis. They charted out the evolution of the solar system in following sequence:

- The Sun had a companion star with which it formed a binary system.
- A third star passed close by the companion of the Sun but it was very far away from theSun itself. It, therefore, did not affect the Sun.
- A tidal filament was formed on the surface of the companion star.
- The gaseous filament was ultimately ejected from the companion star at great distances from the Sun.
- In course of time, matters of the filament condensed to form a system of very closely spaced planets in much the same manners as conceived under gaseous tidal hypothesis.
- Satellites were formed under the mutual gravitational pull of the planets and nucleus of gaseous matters which were formed out of the remaining matters of the filament.

## Merits of the hypothesis

-   The hypothesis is apparently based on the observed facts in the universe about numerous system of binary stars.
-   The hypothesis unlike **Jeans** and **Jaffrey**'s is able to explain the great distances of planets from the Sun.

It is also able to explain the higher angular momentum of the planets in relation to the Sun as they are not believed to have originated from it.

It explains the formation of the satellites which are suggested to have been born due to the mutual gravitational attraction between the planets on which the intruding star could not have any impact.

Despite many positives critique find drawbacks in the hypothesis and which they believe could not be addressed by its proponents. More generally, this hypothesis, according to them, fails to account for the removal of the sun's companion star from its (Sun's) gravitational attraction on the one hand and for the retention of its tidal filaments to have condensed into planets and revolve round the sun, on the other.

## Criticism

■   The hypothesis fails to explain the disappearance of the rest of the Sun's companion star after planets were formed from its ejected filament.

- It is not clear how the Sun's companion could be removed from the solar gravitational pull that made them a binary system.

- It, under the above mentioned condition, is improbable that the planet forming filament remained under the gravitational pull of the Sun.

- If the planets initially were formed close by as the hypothesis suggests then how the present spacing of planets could be achieved? It remains unexplained.

- If the intruding star could affect production of filament on the Sun's companion due to its gravity it should also have the formation due to companion's gravitation pull on its surface. No such suggestion is made in the hypothesis.

**Lyttleton** supporting the hypothesis in 1938 tried to prove through his mathematical calculation that many of the objections against the hypothesis may be removed if it is assumed that:

- All the three bodies i.e. the Sun, its companion and the intruding star had almost equal mass.

- The companion of the Sun in the binary system maintained a mutual distance of 2500 million kilometres.

- It was moving around the Sun with an average speed of about 9 kilometres per second.

- The third star passed by the companion of the Sun at an average speed of about 30 kilometres per second.

- In its journey in intergalactic space it came as close to the companion star as 4 to 6 million kilometres.

He proved that if such conditions are obtained then it is possible that the elliptic orbit of the companion around the Sun was changed into a hyperbolic orbit making it free from mutual gravity. It would facilitate the escape of the companion star into the space. He also proved that due to increasing proximity of the wandering star tidal filaments would be formed on the surface of both- the companion and the intruding star. Condensation of planets from gaseous filaments of both would account for the present interplanetary distances.

## Chapter- 2:2:5: Variants of Binary Star Hypothesis

Attempts to meet the criticisms of the **Russell**'s hypothesis and modifications offered by **Lyttleton,** however, found favour with many scholars as a feasible explanation. Interest in the hypothesis may be understood in later suggestions offered by **Ross Gun** in the form of 'Fission Hypothesis'; the 'Cepheid Hypothesis of **A.C. Banerjee**; and the Nova Hypothesis of **F. Hoyle** and **R.A. Lyttleton**. These hypotheses in their basic assumptions and operational details may generally be considered only to be the variants of the Binary Star hypothesis.

## Chapter- 2:2:5 (1) Fission Hypothesis

Fission hypothesis as propounded by **Ross Gun** in 1939 is basically concerned with the origin of a binary system in the Universe. It is believed that if the evolution of a binary system is properly understood it may account for numerous characteristics of our solar system including the formation of planets, their distances, their angular momentum as well as their rotation and revolution. The basic assumptions behind the hypothesis and mechanism for the evolution of the solar system under **Ross Gun**'s visualization may be outlined in order of their sequence as follows:

- A star experienced increase in its speed of rotation due to decrease in its size caused by its cooling and contraction.
- Stellar matters started to collapse and the process of fission was initiated by increasing thermonuclear reactions in its body.
- It prepared the star to break in two parts.
- At the same time it encountered another fast moving star by chance.
- Being in a state of fission matters from disintegrating star was attracted towards the passing by star.
- A tidal filament from the star on the verge of breaking was ejected under the gravitational pull of the passing star. At the same time the event quickened the process of disintegration of the star.
- A binary system of stars bound by mutual gravity, thus, was created.
- The tidal material drawn from the disintegrating star broke into several parts and started revolving round one part of the broken star representing the Sun.
- Matters of the ejected tidal filament condensed to form planets.

## Merits

- The theory essentially is formulated to address the criticism against the original Binary Star Hypothesis in respect of planetary formation and distribution of angular momentum.

## Criticism

- The hypothesis though accounts reasonably for the distances of the planets and their composition it is the very foundation that there was a chance encounter of the breaking star with another has been considered untenable by most of scholars. This requires to be noted that encounter of stars in the universe itself is a very rare event.

This makes the encounter at the time of fission a very remote possibility and one on which no scientific viewpoint can be developed.

## Chapter - 2:2:5(2) Nova Hypothesis

In 1939 **F. Hoyle**, a British astronomer basing his ideas on existence of novae in the universe and process of their transformation put forth his hypothesis. He believed that the Sun formed a binary system with a much more massive nova star (*Novae are stars which are explosive in nature and are characterized by abrupt and large increase in their brightness during the time of explosion. They experience slow decrease in luminosity after explosion. They are different from the super novae in the sense that they emit much lesser energy in comparison*). Proponents of this hypothesis contended that it was the explosion in the nova that led to the formation of the planets and their satellites of the solar system. The sequence of events as visualized by the proponents may sequentially be put as follows:

- The Sun formed a binary system with a much bigger nova star.
- The nova had a tremendous explosion and emitted large volumes of gaseous matter.
- A large proportion of the expanding gaseous matter came towards the Sun.
- These matters were caught under the gravitational attraction of the Sun.
- They, thus, started to revolve round the Sun.
- In course of time, gaseous matters cooled and condensed into planets.
- Left over gaseous matters in the vicinity caught under the gravitational pulls of the planets condensed to form their satellites and other members of the solar family.
- In the mean time, as a result of explosion the main body of the nova recoiled to a great distance.

- As it receded in inter stellar space its binary relation with the Sun was broken.
- The Sun with the retinue of its planets, their satellites, comets and meteorites formed an independent solar system.

## Merits of the Hypothesis

- The hypothesis is based on the existence of novae and scientific information about their characteristics.
- It provides a plausible explanation for the removal of the companion star of the Sun.
- It is able to account for the great distances between the planets.

It also accounts for the higher angular momentum in the planets as compared to the Sun.

## Criticism

■ The hypothesis is unable to offer satisfactory explanations for the rotation of the planets and the origin of their satellites.

■ It fails to account for a systematic arrangement of the planets according to their size.

■ It is unable to explain the compositional differences between inner and outer planets.

## Chapter - 2:2: 5(3): Cepheid Hypothesis

In 1942 **A.C. Banerji** of India put forth his theory based on the nature of Cepheid (A Cepheid is a type of star which is 5 to 20 times larger in size and much greater in luminosity than the sun. It is characterized by a regular pulsation. The pulsation is believed to be caused by repetitive contraction and expansion in the Cepheid due to continuous collapse of burnt matters and subsequent tremendous heat generation from its subsurface. Radial difference between two maxima may be millions of kilometre). They came to be known to the astronomers since the first star of this type was identified by **John Goodricke** in 1784. According to this theory the birth of the sun and the planets was probably caused by an encounter of a Cepheid with another star. Such an event according to **Banerji** would set a chain of reactions in the Cepheid leading to its partial disintegration. He outlined the course of events prior to the formation of the solar system as follows:

- A star in course of its journey in inter stellar space came close to a Cepheid.

- It led to increase in the rate of pulsation in the Cepheid that made it unstable.

- Instability, thus, induced is believed to have resulted in tidal action and ejection of a large amount of material from the surface of the Cepheid.

- The velocity of the tidal ejection is suggested to be so high that the ejected matters were freed from the gravitational influence of the Cepheid.

- The ejected matters in course of time started condensing to form the sun.

- Rest of the matters condensed to form the planets.

The Sun with 2/5th of the Cepheid matter moved in one direction and the Cepheid in another under the influence of the fleeing star. It freed the Sun from the binary system The following diagrammatic representation of the hypothesis may illustrate the formation of the solar system;

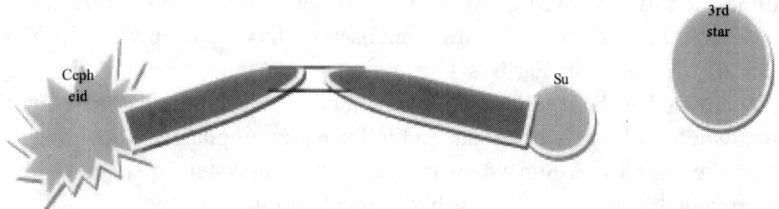

**Fig. 2.4: Evolutionary Process under Cepheid Hypothesis**

## Merits of the Hypothesis

- The hypothesis is constructed upon mathematically proven fact that the Sun has almost 20 times more energy than all the planets combined together. The same proportion is believed to have existed between the Cepheid and the Sun.

- The hypothesis accounts for the revolutionary motion of the planets. According to this hypothesis it is the gravitational pull of the other star which induced the revolution. But it itself receded away into the space leaving them bound with the Sun.

- The hypothesis is able to explain the great inter planetary distances.

## Criticism

- Like other hypotheses invoking binary system the Cepheid hypothesis also fails to account for the rotational motions of the planets and their satellites.

- It likewise is unable to explain the systematic arrangement of the planets and their compositions.

In the light of its failure to address the drawbacks by 1940 the hypotheses invoking binary star system as the cause of the origin of the solar system were more or less rejected by the scientific community.

## Chapter - 2:2:6: Electromagnetic Hypothesis

In 1942, **Hannes Alfven**, a Swedish plasma physicist, proposed an entirely different view point about the origin and evolution of the solar system. Instead of either gravitational attraction or the tidal force of primordial celestial body he invoked electromagnetism to explain the phenomenon. Though he could not develop the hypothesis in full detail it gave a new dimension to the vexed question of the origin and evolution of the solar system. He based his study on some observed facts. It is now established that the Sun like our Earth is surrounded by a magnetic field. Following the knowledge about solar magnetic field he hypothesized different stages that would ultimately culminate in the formation of planets. According to him the primordial Sun when it was without a system of planets was surrounded by nebular dust. It had a much stronger magnetic field then than its gravitational force. Such a situation is believed by him to have set a sequence of changes leading to the formation of a unique system that is our solar system. He assumed that:

- The Sun then had a much faster rotation as compared to the present (it is now established that rotation of almost all the members of the solar system has been slowing down).

- Though the gravitational force of the Sun was lesser than the force of its magnetic field it still could attract the surrounding nebular atoms towards it.

- While encountering the solar magnetic field nebular atoms were ionized.

- With passing time the Sun was surrounded by clouds of these ionized atoms.

- Clouds of ionized atoms spread as far as the farthest planet of the present solar system.

- Due to solar rotation the ionized atoms collected themselves about the equatorial plain of the Sun. In tune with the law of motion for charged particles the cloud disc would extend up to distances of Jupiter and Saturn.
- Ionized atoms of the nebular cloud would start to revolve round the Sun along its equatorial plain.
- Revolution of these atomic matters around the sun is believed to have resulted in slowing down of the Sun's rotation.
- At the same time, due to gradual cooling, the atomic matters would start condensing to form different planets.
- The magnetic field of the planets thus consolidated would attract left over ionized atoms which after condensation would form their satellites.

## Merits of the Hypothesis

- The hypothesis is one that suggests an alternative mechanism of the evolution of the solar system sidetracking the role of both the gravity of the parent bodies or it's off shoot in the form of tidal effect as adopted in most of the preceding hypotheses.
- It is based on facts like electromagnetic fields around the Sun and the Earth and their impact on ionization of inter galactic atoms.
- It is capable of providing a scientific explanation for the origin and evolution of the solar system.

## Criticism

- The hypothesis has not been developed in its full detail about the solar system.
- It also fails to account for the formation of inner planets for they would be exposed to a very strong solar magnetic field. How ionization of nebular atoms will be consolidated to form them is not clear.
- If the ionized nebular matters collected along the equatorial plain of the Sun to a maximum distance of the Saturn then formation of Uranus and Neptune is not explained.

Whatsoever may be its limitations, the hypothesis is accredited with infusing new dimension in the field of cosmogony and to the study of the solar system.

## Chapter - 2:2:7 Interstellar Dust Hypothesis

Interstellar Dust Hypothesis was proposed by **Otto Schmidt**, a Soviet scientist, in 1943. It is inspired by the propositions made by **Kant** and **Laplace** in late 18th century. It, in a way, may be said to have revived interests in Nebular Hypotheses with a suggestion of a different mechanism. The view incorporates in it the verifiable properties of the intergalactic space. It is based on the fact that besides stars a large amount of gases and dust particles are still found scattered in the universe. He believed that it was the interaction between the stellar masses and widely spread cosmic gases and dust that guided the course of evolution in intergalactic space. Similar interaction is suggested by the hypothesis to have led to the evolution of our solar system. Schmidt systematically outlined the course leading to the evolution of the planetary system around the Sun as below:

- The sun was already in existence and attracted some gases and dust particles due to its gravitational attraction.

- These particles formed a sheet of cloud and dust revolving around the sun in an anti clock wise direction.

- In the initial stage the gas molecules and dust particles of different sizes revolved similarly round the sun in an unorganised manner due to density variations and gravitational anomalies.

- This is believed to have been causing collision of particles leading slowly to their accretion and making them relatively heavier.

- In course of time, however, the heavier particles, as would be natural, got concentrated near the bottom of the cloud whereas the lighter particles formed the upper layers of the cloud.

- The cloud heap gradually attained the shape of a vast saucer in which mutual collision of particles further consolidated and enlarged them.

- Gradually the particles within the cloud underwent rearrangement consequent upon temperature variations/differentials in respect of their location within the cloud sheet. Heavier particles being collected at the bottom of the cloud heap and lighter ones horizontally more spread over it.

- Consolidated particles in the dust cloud formed embryos of planets.

- Further consolidation of matters of dust clouds, in course of time, gave them the form of asteroids.

- These asteroids started revolving around the sun within the dust disk in the direction of the latter.

- The asteroids grew in size by collecting and absorbing the scattered matter within the cloud and developed their own field of gravity.

- They, thus, turned into planets.

- Even after the formation of the planets enough materials remained scattered in the nebular cloud in attenuated form. They started to revolve round the planets. In course of time these materials further consolidated and enlarged to form the satellites.

## Merits of the Hypothesis

- The theory is found to resolve a very vexed problem about the distribution of angular momentum between the Sun and the planets. Because planets are believed to have been formed independently from stellar dusts and gases and not from the Sun, it is not unnatural that the planets are having greater angular momentum than the Sun. The difference is attributed to the angular velocities of gas and dust particles at the time of their condensation and consolidation.

- The hypothesis is also able to explain the difference in composition of the inner and outer planets. It is suggested that the inner side of the saucer due to proximity with the Sun was hotter. It would facilitate concentration of heavier elements like silicon, iron, and aluminium etc. of which inner planets are composed. On the other hand, the outer part of the saucer was cooler and allowed condensation of lighter elements like hydrogen, helium, methane, ammonia etc. to form the outer planets.

- Distances between different planets in the solar system, it is believed, is in accordance with the statistical laws which suggests that bodies of dissimilar size and velocity is more likely to consolidate at certain fixed distances from the Sun while revolving round it.

- The hypothesis has also been able to address the question about circular orbit of the planets around the Sun. It has been attributed to average motion of mutually colliding particles which is believed to have led to similar orbital path and movement in the same direction.

- Thus, the interstellar dust hypothesis seems to solve a number of problems such as the shape of the orbits of the planets, their distribution by mass and densities and over all the difference between angular momentum of the planets and the Sun.

## Criticism

Astrophysicists, however, point out to certain controversial points in this hypothesis. Important criticisms are listed below:-

■ It is pointed out that if the gravity of the Sun was responsible for inducing motion of gas and dust particles around it then why the gravitational force of the Sun has been considered inadequate to capture the gas and dust particle scattered in the universe. No satisfactory explanation is provided by the pioneer of the hypothesis.

■ The assumption that the planets were formed out of asteroids is also considered by many as improbable. They, on the contrary believe that known asteroids were formed due to disintegration of planets.

## Chapter - 2:2:8 Nebular Cloud Hypothesis

**Karl Von Weizsacker**, a German physicist presented this hypothesis in 1943. He attributed the origin of the solar system to condensation of fine particles of dense interstellar clouds of gases and dust of a nebula surrounding the sun. In his basic assumptions he has been remarkably close to that of **Otto Schmidt**. He like Schmidt also believed that the interstellar space was filled with mixture of gases and fine dust particles. But the similarity of the two ends there. **Weizsacker**'s proposition about the mechanism of the evolution of the solar system is substantially different from that of his contemporary. He believed that the chemical composition of the sun and the other stars was similar to that of the nebular cloud. It implies that the Sun either was formed due to condensation of parts of this matter or it might have entered the cloud from beyond. In either case it is suggested that a giant cloud of the interstellar matter surrounded the Sun. He then elaborates the stages in cloud-Sun interactions and its resultants in the following manner:

• This nebular cloud was composed mainly of hydrogen, helium and small but significantly higher density dust particles of iron oxides, compounds of silicon, water droplets and ice crystals.

• This cloud started to revolve like a ring round the sun under its gravitational influence.

- The revolving ring of cloud developed numerous circular eddies (vortex) due to gravity anomaly of matters within it. It would cause accumulation of heavier matters at the base of vortex.

- It is suggested by **Weizasacker** that the nebular cloud extended approximately up to present diameter of the solar system having a thickness of about 300 to 450 kilometres.

- During the course of initial unorganized revolution dust molecules within the nebular cloud started to collide with each other and aggregated wherever and whenever possible (*aggregation by collision being a function of velocity of colliding matters*). In this manner, it is suggested that large lumps of matter were formed.

- The process of collision and aggregation is believed to have quickened as growing lumps of matter would acquire sufficient gravitational force to attract matters in their vicinity.

- In about 100 million years, it is suggested that these lumps grew in size to ultimately form the planets.

- As the size of lumps of aggregated matter grew in size, proportion of dust particles in the nebular cloud correspondingly decreased.

- It is also suggested that the size of the lumps of matter or planetoids or the planets would depend on their optimum distances from the Sun that is believed to have induced revolutionary motion in the nebular cloud. Thus, planets of various sizes would be formed firstly in increasing order moving away from the Sun and then in decreasing order farther away from the Sun.

- At the same time, in-falls of nebular matters during their growth made the planets hot. This would also lead to their phase change and rearrangement resulting in accumulation of high density matters at the core to be surrounded by lower density matters.

- Once the size of planets stabilized the planets started losing heat due to radiation.

- Gradually they developed a solid crust.

- Interplanetary remains of the nebular cloud consolidated as satellites of different planets in much the same fashion in which planets are believed to have been formed around the Sun.

## Merits of the Hypothesis

- This hypothesis seems to be highly probable and is quite logical. Even today there are nebular clouds in abundance in space.

- The passing of the sun into a cloud of dust and gas, or diffused nebula in the constellation is not an assumption any more. There are numerous stars which are found to be in similar situation in contemporary universe.

- Distances of planets and their sizes as well as the formation of satellites are scientifically accounted for. It is found that the distance of planets from the sun is roughly twice the distance of the next inner planet. It is in accordance with the postulates of the hypothesis.

- The difference of angular momentum between the Sun and the planets is explained.

- The hypothesis reveals a possibility of more systems like ours in the galactic space. Now it is proved that t many planetary systems in the Universe do exist.

- The hypothesis though has been successful in explaining many characteristics of the solar system there still remain some unexplained and unaddressed facts for which the hypothesis is being criticized.

## Criticism

- How the planets attained their rotation generally in one direction is not very clear from the hypothesis. In fact random and unorganized collision of particles would give different direction to the motion. It may, however, be possible that the gravity of the Sun dictated the rotational motion of the inner planets. It would not be possible for the outer planets to be affected in the same manner because there is an inverse relationship between distance and gravitational attraction.

- The hypothesis is unable to explain why Mercury one of the inner planets and the Uranus one of the outer planets have a retrograde rotation when others planets are having prograde rotation. Why many satellites of Jupiter, not all, are having retrograde rotation.

- The hypothesis also fails to explain the varying inclinations of the axis of the planets on their orbital path.

- The turbulent eddies presumed to be the cause for the formation of planets is not found to be as regular in size and arrangement as suggested in the hypothesis.

- The hypotheses put forth by Otto Schmidt and Weizsacker have, however, revived the nebular hypothesis of Kant and Laplace as a possible explanation for the origin of our solar system with more positives than their drawbacks.

## Chapter - 2:2:9: Protoplanet Hypothesis

Protoplanet hypothesis as put forth by **Gerald Kuiper**, an American astronomer in 1951, may be considered a modification of **Weizsacker** views to start with. However, while doing so he appears to have suggested a mechanism of planetary evolution which is substantially different from that of his predecessor. Thus, it is better to consider his views as a new hypothesis. He unlike Schmidt and **Weizsacker**, introduced a new dimension to the origin of the solar system. He believes that the primordial nebula composed mainly of hydrogen, helium and a very low proportion of heavier elements, probably not more than 2 % all together, started to condense first in its outer sphere. At the same time nebular matters were collapsing around its central core under its own gravity setting in a thermonuclear reaction to make it more luminous. He, thus, assumes that planetary bodies which are less massive formed before the central core of the nebula evolved as the Sun. In other words, the planets are older than the Sun. This is considered to be a significant departure from all the earlier hypotheses. With this basic assumption he visualized the evolution of the solar system as outlined below:

- There existed a rotating nebula in the interstellar space composed mostly of hydrogen, helium and a very small proportion of heavier elements (heavier elements could not be more than 2% of the total volume of the nebula. It implies that there was gravity anomaly within the nebula).

- The direction of rotation of the original nebula was anticlockwise.

- As the nebula cooled, it contracted. It also developed some core areas within the outer nebular sphere having a collection of solid particles. As a result continuous nebular cloud got divided into a number of separate clouds or what **kuiper** considered being proto planets.

- Each cloud of proto planets, thus, is believed to have originated as a collection of solid particles at the centre surrounded by extensive envelope of lighter nebular matters.

- The proto planets were of different size but were bigger than the present planets.

- As the proto planets contracted, the satellites were formed close to the planets from the nebular matters surrounding them through a similar process.

- While the planets and satellites were forming, the nebula under the gravitational pull of its central core was also collapsing and condensing slowly. The central core was, thus, developing into a star-the sun.

- At this stage due to the gravitational pull proto planets were elongated with their long axis pointing towards the nebular core (prospective Sun). This led them to rotate in the same direction i.e. anti clockwise in which they were revolving around the central core as a part of the nebular cloud.

- Simultaneously thermonuclear reactions in the central core generated intense heat and increasingly more luminosity in the process of its transformation to Sun.

- It caused intense radiation and solar wind driving the ejected particles into space.

- The lighter gases of the proto planets were also driven away similarly and the heavier parts of the proto planets assumed the form of the planets.

- Complete removal of the lighter gases from planets closer to the sun resulted in their higher density. The planets at a greater distance retained part of these lighter gases and thus have a lower density. The two largest planets, the Jupiter and the Saturn have the lowest density.

- As the mass and the size of the proto planets decreased, their gravitational attraction also decreased proportionately. As a result their satellites also moved a little farther away from the position they were formed.

- Decrease in the mass and size of planets from those of proto planets also resulted in the increase of speed of their rotation.

## Merits of the Hypothesis

The hypothesis is able to explain most of the features of our solar system which could not be explained by earlier hypotheses.

- It is found to have successfully accounted for the shape size and distances of the planets.

- It provides a logical explanation of the distribution of angular momentum between the planets and the Sun as they are considered to have been formed separately at different times.
- It attempts to explain the direction of rotation and revolution of the planets and their axial inclinations.
- It explains the compositional variations of the planets.
- It provides a logical framework of the distribution of satellites.
- It explains the interior composition of planets.

Empirical evidences produced in support of the hypothesis particularly with regard to temperature variations and composition as well as dimensions of terrestrial and Jovian planets made scholars like **R.J. Ordway** and the Physics noble laureate **Harold C. Urey** to support **Kuiper's** view about the origin of our solar system.

## Criticism

The hypothesis, however, is not without some shortcomings and certain explanations are not found to be plausible under existing knowledge of physical laws.

- One major objection against the hypothesis is in respect of the suggested collapse of proto planetary matters to form planets. Physicists suggest that in order to induce collapse of the proto planets their mass should be almost equal to the mass of the Sun. It is now known that the Sun presently has more than 99% of the total mass of the solar system. Thus collapse of proto planetary cloud to form planets would not be possible.
- In the light of the above the formation of the inner planets is difficult to be explained because a giant proto planet would not be stable at such a close distance due to tidal force of the Sun.
- To many critics the hypothesis is unable to explain fully the chemical composition of the planets which are distinct in inner and outer planets.

## Chapter - 2:2:10 Collision-Accretion Hypothesis

It is obvious from the above mentioned hypotheses that there has been an increasingly better understanding and explanation about the mechanism responsible for the evolution of solar system. However, mysteries of the universe have not yet been comprehended fully despite the fact that space expeditions and round the clock functioning space observatories equipped with ever powerful instruments in different parts of the world have been providing more and more specific data from the interstellar space. With greater understanding of the space new propositions are being made many of which not yet crystallized. However, based on geophysical data on the terrestrial planets **Safronov**, a Soviet astronomer, propounded his collision- accretion hypothesis in 1971. The hypothesis suggests gradual accumulation of increasingly larger solid matters as the basic cause of the formation of planets within a nebular body. Safronov believed that the planets were formed in three distinct stages in which mechanism of their formation varied. In proposing his hypothesis **Safronov** assumed that in the initial stages of the evolution of the planetary system the Sun was enveloped by a nebular mass composed mostly of relatively homogeneous matters. He then identifies three stages in the evolution of the planets under different time frames.

The first phase, he believes, was characterized by condensation of matters initiated by smallscale gravitational instabilities in the solid dust grains present in the nebula. Such condensation is believed to have formed small planetoids* of not more than 5 kilometres in diameter. Due to varied nature of dust particles and quicker consolidation this stage is suggested to have lasted only for about 1000 years. These planetoids revolved round the Sun in orbits as suggested by **Kepler** in 15th century. However, their orbits due to gravity anomaly were disorganized. This is suggested to have resulted in destruction of the planetoids due to collision. Alternatively, planetoids passing each other without collision would experience eccentricity in the course of their orbital path caused by the effect of mutual gravity.

The second or intermediate phase according to **Safronov**, was characterized by the growth of planetoids under two conflicting yet simultaneously operating processes - (i) a very rapid destruction upon collision of planetoids consequent upon their unorganized orbital motion, and (ii) a slow accumulation of matters to enlarge the size of planetoids[1]. This phase is suggested to be characterized by destruction of large number of planetoids due to mutual collision. Some planetoids, however, could accumulate matters due to accretion on encounter and grew in size to about 1000 kilometres. Safronov identified this phase also as e**the phase of embryo planets'**.

In the third and final phase, planets are believed to have grown to their present size having mutual gravitational attraction between the embryo planets. Phase II and III of planet formation together is believed to have spanned over 100 million years.

It needs to be emphasized that if the planetoids have a very close encounter with a very high approach velocity a collision is more likely to take place destroying the planetoids. On the other hand, if approach velocity is low the planetoids are likely to fuse together under their mutual gravitation creating a larger and bigger body. Once the embryo of planets attain an optimum size of about 1000 to 2000 kilometres the growth in their size would not depend on chance encounter of the planetoids. Rather they would enlarge in size by attracting matters from far beyond their diameter due to their enhanced gravitational force. Therefore, any planetoid entering this zone would be absorbed by the embryo planets till they grow to a certain size. Also with reduction in chance encounters of planetoids after embryo planets attain a certain size their paths of movement would be more organized and their orbits would be more independent and circular.

## Merits of the Hypothesis

The hypothesis successfully addresses the questions which many of the earlier propositions failed to explain. The hypothesis is able to explain:-

- The number of planets and their size,
- Their distances and their mass,
- their rotation

## Criticism

- The basic assumption of **Safronov** like most of the preceding hypotheses of similar kind that the Sun predates the planets has been challenged. Many scientists believe that all the members of the solar family were formed simultaneously It is, however; obvious from the evaluation of different hypotheses that still there is no full comprehension of the mechanism which might have propelled the formation of our solar system. But it is increasingly believed that there have been orderly and systematic natural processes that created our solar system and not any accidental catastrophe. The foregoing brief overview of the various theories trying to explain the origin of the earth indicates that the basic assumptions of early scholars like Kant and Laplace are once again finding a support amongst the modern theorists. This, however, must be realized that till early 90s only our solar planetary system was available to the scientists as a model of study. Discovery of extra solar planetary system starting with 51 **Pegasi** in 1995 has provided the scientists with many more planetary models and which are under intensive scrutiny. The results obtained there from may necessitate re-evaluation of contemporary theories of planetary formation.

## Reference

1. **Audouze, J. and G, Israel** (Eds) (1986): The Cambridge Atlas of Astronomy, Cambridge University, Cambridge.

2. **Christian, D.** (2011): Maps of time: an introduction to big history; University of California Press, Berkeley.

3. **Darling, D**. (1999): Jeans-Jeffreys tidal hypothesis, Doctoral thesis, University of York.

4. **Dayal, P.** (1990): A Text Book of Geomorphology; Shukla Book Depot; Patna.

5. **Hawking, S.** (2011): The Theory of Everything: The Origin and Fate of the Universe; Jaico Pub. House; Ahmedabad;

6. **Smart, W.**M. (1959): The Origin of the Earth; Longman; London.

7. **Steers, J.A.** (1964): The Unstable Earth; Methuen & Co. Ltd. London.

8. **Wooldridge, S.W. and R.S. Morgan** (1948): The Physical Basis of Geography; Longman, London Read more: solar system: Origin of the Solar System http://www.infoplease.com/encyclopedia/science/solar-system origin-solarsystem.html#ixzz3ZjIOoR6d

## Note

1. **Planetoid** Planetoids are small celestial bodies that move around the sun.

## Review Questions

1. Bring out the major characteristics of our Solar system.

2. What are the major approaches to the study of the origin of our solar system.

3. Assess the theories about the origin of our solar system propagated after 1940.

4. Laplacian theory about the origin of planets is being revisited. Evaluate the statement in the light of recent theories.

# Chapter - 3

# The Earth as a System

## Earth and Sun

## Objectives

Main objective of this chapter is to make its readers understand the basic relationship that exists between Earth and the Sun. It also attempts to explain the operational mechanisms that propel the system differentially in different segments of our living planet and its understanding in historical perspective. It particularly emphasises on energy dynamics and factors affecting the Earth system as well as practical use of their understanding that impacts routine functions of its inhabitants. It is also considered crucial to understand the uniqueness of this living planet. This chapter, thus, undertakes the study in respect of:

- Shape and Size of the Earth.
- Movements of the Earth- Rotation and Revolution.
- Earth's orientation with the Sun- Seasons and length of days.
- Referencing - Latitude and Longitude

## Chapter - 3:1 Shape and Size of the Earth

The Earth, so far the only known living entity in interstellar space is the biggest of the terrestrial planets in our solar system. But questions about its shape and size perplexed the people since ancient times. Greek scholar like **Thales** of Miletus considered the Earth to be a flat disc floating over water like liquid. **Pythagoras** (580 –500 B.C.) followed by **Aristotle's (384 BC – 322 BC)** observation was probably the first person to prove mathematically that the Earth was spherical and not flat. However, it was not till the first century A.D. that the view found greater acceptability particularly amongst the educated ones. In fact, many of the voyages during the **Age of Discovery** and **Renaissance** were successfully carried out on this premise. Geodetic surveys till last quarter of the 17th century were also conducted with the assumption that the Earth was a **spheroid**. The belief that the Earth was spherical was challenged by **Isaac Newton** in 1687. Based on knowledge of faster rotational motion on equator and slower motion at higher latitudes he proved that the Earth could not be perfectly spherical. If at all it could be an **oblate** (flattened) **ellipsoid**. With development in space technology the shape of the Earth now has been determined. There is no other shape that can resemble that of the Earth. It is this understanding of the shape of the Earth that it has been given an independent name-**geoid** implying that the Earth's shape is like that of the Earth only.

Similarly, the knowledge about the size of the Earth has also undergone substantial change from the ancient times. Scholars have always been trying to establish the exact dimensions of this living planet. Despite limited information and limited access to precision technology many of the ancient geographers led by **Aristotle** believed Earth to have a spherical shape. But the first scientific attempts to measure the spherical dimension of the Earth was made by **Erosthenese** in 247 B.C. The polar circumference according to his calculation would be about 250,000 **stadia** (a unit for measurement of length in Greece during those days. However, stadia length used to vary in different parts of Greece.) One commonly used stadia, **Gabler** et al (2009:79) believe, roughly equalled to 0.157 Kilometres

---

**How Erosthenese measured the Polar circumference:-**

He measured the inclination of Sun's rays by using an obelisk (a perpendicular column) of similar height at Alexandria and Syene about 5000 **stadia** apart and ostensibly on the same meridian. He selected noon of 21st June when the Sun was believed to be at its northern most point (the day of summer solstice) for his experiment to measure the shadow of the obelisk at the two points. The readings that he got were $0^0$ at Syene and $7.2^0$ at Alexandria. The angular difference between the two points of $7.2^0$ represented almost $1/50^{th}$ of $360^0$, the circumference of a sphere. Multiplying the distance between the two points with 50 he concluded that the circumference of the Earth was 250, 000 stadia.

---

or 515 ft[1]. Erosthenese's finding would translate in terms of this measurement to about 39,250 kilometres (24,388 mi). His calculation is found to be remarkably close to the present reading of the Earth's polar circumference of 40,008 km (24,860 mi). In post Newtonian period now equatorial circumference has been calculated to be 40,075 km (24,902 mi) with a diameter of 12,758 kilometres (7927 mi), while from pole to pole it is 12,714 kilometres (7900 mi). Equatorial diameter, thus, is about 44 kilometres bigger than its polar diameter.

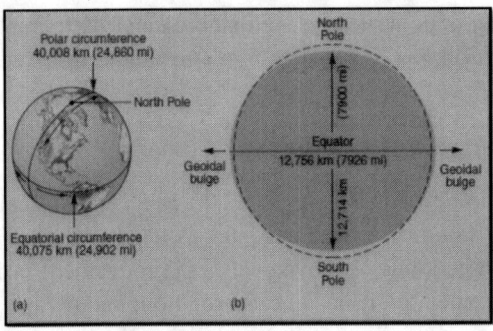

*Fig. 3.1: Size of the Earth*

The shape of the Earth can be mathematically represented as an ellipsoid defined by:

– Semi-major axis = equatorial radius = **a**

– Semi-minor axis = polar radius = **c**

– Flattening (the relationship between equatorial and polar radius): **f = (a-c)/a**

– Eccentricity: e2= 2f-f2 (After Calais Purdue University-EAS Department Civil 3273 ecalais@purdue.edu

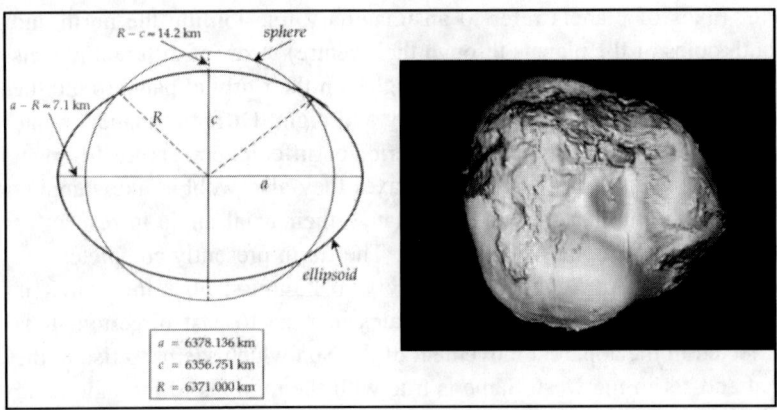

(After Calais Purdue University -EAS Department Civil 3273 ecalais@purdue.edu)

*Fig. 3.2: Shape of the Earth.*

However, the difference between a perfect spherical Earth and the actual shape of the Earth is very minor. The deviation is found to be only 1/300. So the Earth for general practical purposes may be considered spherical in shape. Only for very precise measurements as required in the fields of navigation, surveying, aeronautics and cartography the actual shape of the Earth has to be taken into consideration.

## Chapter - 3:2: Movements of The Earth- Rotation, Revolution and Axial Tilt

The relationship between the Earth and the Sun is intrinsically associated with different motions of the Earth and its axial tilt. The Earth may be said to have three kinds of movements - (1) movement in intra galactic space; (2) movement on its axis i.e. its rotation; and (3) movement on its orbital path around the Sun i.e. its revolution. Of these, movement of the Earth in intra galactic space refers to its movement around the centre of our Milky Way and which it undertakes collectively with other members of the solar system. It is believed to have a very little or no effect on the Earth environment. It is, however, significant to astronomers and interstellar researches including the operational details of inter planetary/ inter galactic spaceships. Physical Geographers are more concerned with the two other motions which directly impinge on environment of the Earth. They are (i) the rotation, and (ii) the revolution of the Earth.

## Chapter - 3:2:1: Rotation

Rotational movement refers to the spinning motion of planets on their axes (axes of planets refer to an imaginary line joining the north and south poles of the planets through their centre). Axes of different planets are generally inclined at different angles on their orbital path around the Sun. It is this motion that causes day and night. Different planets rotate at different velocity. Thus, rotation period of different planets are different. In the process of spinning on their axes they also wobble like spinning tops. This wobbling motion may change their axial angle in respect of their plane of ecliptic (see axial tilt). The Earth presently completes one rotation in about 24 hours (precisely in 23hours, 56 minutes). It is equivalent to one Earth day. It rotates in west to east direction. It is reflected in the apparent movement of the Sun which seems to rise in the east and set in the west. Same is true with the motions of the moon and other stars in the universe. Viewed from the North Pole the rotation of the Earth would appear to be anti clock wise. It is the rotation of the Earth that defines the movement of day light and atmospheric and oceanic circulation. However, the rotational velocity of the Earth like the Sun and other planets is decreasing. It means that its sidereal period of rotation is increasing. It is suggested that some 150 million years ago the sidereal period of rotation of the Earth was only about 22 hours. It must be noted that due to slowing of rotational movement there has been readjustment of **Universal time** at a rate of a little less than 1 second almost every two years since 1972. Most recently the world watches were readjusted by one **leap second** on 30th June, 2015 at 23:59:60 UTC. Since 1972 the earth rotation has slowed down by 26 seconds. It is considered to be a very rapid rate and shows that rotation of our planet is increasingly decreasing at a faster rate. This naturally is going to reflect on the length of days and exposure to solar radiation. It is now estimated that after 150 millions years from now sidereal period of rotation will increase by three hours to about 27 hours. Another characteristic of the Earth's rotation is the variations in its **linear velocity** at different latitudes in relation to the motion at the equator. It must be noted that the speed of rotation of the Earth is 830 kilometres (519 mi) per hour. As the Earth is spherical all points on it rotate in 24 hours (150 in one hour). In other words, the movement of 3600 is completed in 24 hours. It means that the angular velocity at all points on the Earth is same. But the time to cover linear distances on different latitudes is bound to vary because their lengths are different. It should be understood that the rotation on the equator has to cover a length of 40,075 km (24,902 mi). Rotational movement near the equator, thus, covers a linear distance of about 1660 kilometres (1038 mi) per hour whereas at 600 latitudes it has to cover only half the distance than that at the equator. Both the poles represented by point does not have linear dimension. Therefore, no linear distance is covered there in a rotation of 24 hours.

However, one does not realize the differential speed of rotation because of the fact that

(a) Angular speed of rotation is same at all points on the Earth,

(b) The atmosphere being the part of the Earth system also rotates at the same velocity along with it, and

(c) There is no other reference point available to experience the linear motion of rotation.

## Impact of rotation

Rotation impacts the Earth and its inhabitants in three significant ways- (i) Rotational axis provides the basis on which geographic grids are prepared for referencing; (ii) It influences Earth's exposure to solar illumination and helps in measurement and division of time; and (iii) It significantly influences the processes that regulate physical as well as biotic attributes of the Earth.

❑   Rotation induces bulge at equator whereas its axis that joins the poles through the centre of the Earth provide almost fixed points for reference to any segment on the Earth's surface. Preparation of reference grid that involves lines of latitude and longitude, thus, is made possible by rotation.

❑   Earth's day length is determined by rotation.

❑   In the process of rotation each longitude experiences sun rise, noon and evening but at different times. However, half of the Earth is always exposed to the solar light denoting day. Other half likewise always remains under shadow denoting night. The ever changing line that separates the lighted portion from that of darker side is known as **circle of Illumination.**

❑   Rotation induces coriolis effect by which the atmospheric and oceanic circulation on the Earth are found to be moving in clock wise direction in northern hemisphere and counter clockwise in southern hemisphere.

❑   It causes temperature variation on daily basis as day and night temperatures are generally different.

❑   Not only this, exposure to sun light during day time allows photosynthesis in plants and which is considered real foundational resources for this living planet.

❑   It leads to the differentiation between day and nocturnal organisms.

❑   The rhythmic occurrence of sea tides is intrinsically related with rotation.

## Chapter - 3:2:2: Revolution

The Earth besides having a rotational motion on its axis also moves in counter clock wise direction around the Sun like most of the planet in our solar system. This movement around the Sun is known as 'revolution'. The Earth revolves round the Sun on a definite path. It is known as **orbit**. The rotating Earth moves in its orbit with a speed of 107, 280 km/h (66,660 m/h). This speed and the Earth distance from the Sun determine the time that the Earth takes to revolve round the Sun. It also determines the length of seasons on Earth. It takes about 365 earth rotation (*365.2422 rotation to be exact =365.24*) to make a complete revolution around the Sun. The period is measured in terms of **tropical year**. A tropical year has a reference to the elapsed time between two equinoxes when the Sun is believed to be directly over head equator. Excess time over 365 days (rotation) equals to almost a day in four years when a day is added to the year. The year popularly is known as **leap year**. The leap year is adjusted as the multiple of four years. In other words, any given year which can exactly be divided by four is a leap year and when the month of February has 29 days instead of normal 28 days.

The orbit of the Earth around the Sun varies from circular to elliptical. This orbital path of the Earth is referred to as the '**plane of the ecliptic**'. Average distance of the Earth from the Sun is about 150 million kilometres (93 million mi) on this elliptical path. However, Sun (**focus of ellipse**) is not located exactly at the centre of the elliptical orbit. During the period of its revolution around the Sun, thus, two opposite situations arise- (i) when the Earth is relatively closer to the Sun at a distance of about 147,500,000 km (91,500,000 mi), and (ii) when the Earth is relatively farther from the Sun at a distance of about 152,500,000 km (94,500,000 mi). Generally the Earth is closest to the Sun around 3rd January. This situation is known as "**perihelion**". When the Earth is farthest from the Sun it is known as "**aphelion**". Aphelion generally occurs around 4th July and coincides with northern summer. However, a variation in distance of about 5,000,000 km (difference only of about 3.5%) between perihelion and aphelion is rather insignificant in terms of solar energy receipt on Earth. Therefore, they have little bearing on seasons on Earth.

However, plane of ecliptic which is supposed to exist in line with the solar equator is very often disturbed by planetary perturbations. Such perturbations may lead to positive or negative angular changes of the ecliptic in relation to the solar equator. This is known as obliquity of the ecliptic. Presently, the obliquity of Earth's plane of ecliptic is decreasing at a rate of 47" in a century.

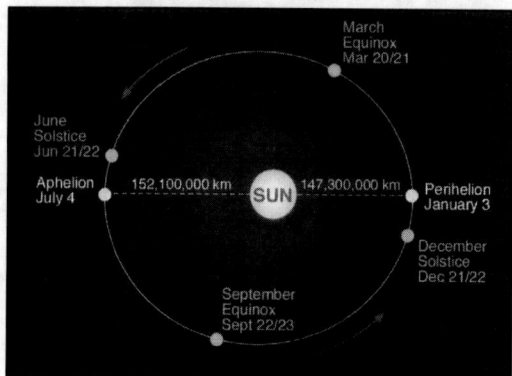

**Fig. 3.3: Position of the Earth on its Orbit**

## Chapter-3:2:3: Axial Tilt

Axial tilt of the Earth also referred to as its obliquity, is the angle that it creates between its rotational axis and its orbital axis. In other words, it refers to the angle that exists between its equatorial plane and orbital plane. The Earth rotates on its ecliptic plane not perpendicularly but with a tilted angle. Axis of the Earth is inclined to its orbital path at an average angle of 23.50. This obliquity or the inclination or the tilt of the axis of the Earth is maintained throughout its journey on its orbital path. It means that the Earth's axis remains parallel to its former positions during the course of its revolution around the Sun. This axial characteristic is known as **parallelism.** However, Earth's inclination is not a permanently fixed angle. It undergoes changes over a period of 42,000 years. This axis is recorded to have been fluctuating between angles of 220 and 250. Presently, the inclination of the Earth's axis from perpendicular is 23.40. During this period the Earth also changes its **axial direction** in a cycle of 25,800 years. The cycle may vary between a period of 20,800 and 29,000 years.

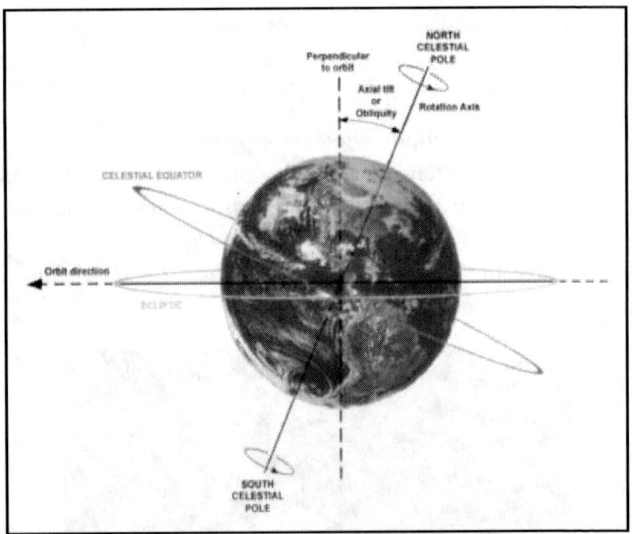

*Fig. 3.4: Earth's Obliquity*

A change in the rotation axis of the Earth impinges on the receipt of solar energy and the rhythm of seasons. Similarly, a change in the rate of spin of the Earth is expected to have far reaching impacts in terms of solar radiation and growing periods of vegetation.

## Chapter - 3:2:4: Earth's Orientation With the Sun- Seasons and Length of Days

The Sun being the power house of energy is the most important contributor to the survival of the Earth as living planet. Its shape, its rotation, its revolution and its axial tilt together determine its orientation with the Sun and its energy input to the Earth. They together bring about changes with changing orientation with the Sun. As our planet moves around the Sun on the plane of its ecliptic, its position in relation to the Sun perpetually changes. Its axial tilt affects the surface receipt of solar energy. It is this tilt combined with rotational and revolutionary motions of the Earth that exposes northern half of it progressively more to the Sun after March equinox. It marks the start of summer season in northern hemisphere with increasingly longer duration of days till June solstice and their gradual reduction of duration till next solstice in December. During this period southern hemisphere experiences changes opposite to that of the northern hemisphere. However, due to a tilt of only 23.50 the Sun can shine over head maximum twice in a cycle of one year between 23.50 north and south latitudes. It implies that any place in between these two latitudes has the altitude of the Sun perpendicularly above it twice in a year with the exception of the two defining latitudes themselves. On these latitudes the Sun shines overhead only once in a year. When the Solar radiation strikes any place at an angle of 900 it is most intense. Intensity of solar radiation north or south of it is relatively lesser. It is because of the fact that higher angular incidence of radiation covers smaller areas. More inclined solar rays cover proportionately larger area reducing the intensity. The point where the Sun rays with highest angle strike the Earth perpendicularly and most intensely is called **sub solar point**.

### Chapter - 3:2:4:1: Equinox and Solstice:

Sub solar points move apparently between 23.50 north and south in annual cycle of Earth's motion around the Sun. They, thus, provide a basis for identifying certain important latitudes between which they move and enable division of the Earth surface into specific thermal zones. It has been noted since long that around 21st March and 22nd September the sub solar point is located on equator. On these days the circle of illumination connects the North Pole with the South Pole perfectly and all parts of our planet has equal duration of day and night (12 hours each) in a cycle of 24 hours. The position of sub solar point on these days is highlighted by the term **equinox**.

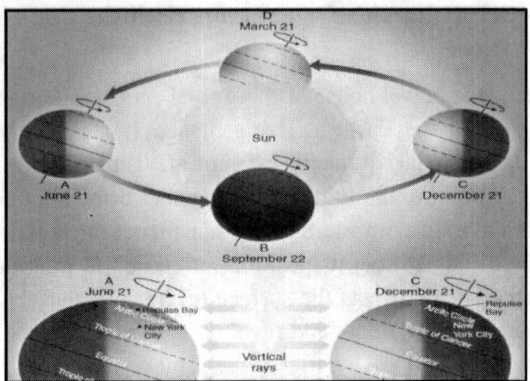

*Fig. 3.5: Cycle of Earth's motion around the Sun*

Similarly, when the sub solar point is located either 23.50 north or south of equator it is denoted by the term solstice marking the occurrence of longest day in respective hemisphere.

The phenomenon is found to occur around 21st June in northern hemisphere and 22nd December in southern hemisphere.

## Chapter - 3:2:4:2: Seasons and Length of Days

Seasons and length of days are very closely associated. It is the duration of exposure to the Sun that impinges on the periodicity and intensity of phenomena that an area experiences as weather conditions. Seasons, thus, may be defined as periods when weather conditions, temperature and length of days over an area of the Earth are on an average similar. Globally four seasons are generally identified as spring, summer, fall (autumn) and winter. They on an average have duration of three months.

Seasons and length of days are intrinsically associated with Earth's orientation with the Sun. The Earth's orientation with the Sun, as is obvious from the earlier discussion, is impacted by its rotation, its axial tilt and its progress on orbit around the Sun. These factors expose the surface of the Earth differentially to the Sun in respect of its angle and duration of exposure to it in a cycle of one year. As a result the receipt of radiant energy of the Sun (**insolation** the *main source of energy on our planet)* on Earth varies over a period of one year. As incoming solar radiation is affected by our planet's changing geometrical relationship with the Sun, it is affected by the variable periodicity of factors like (1) angle of its incidence, (2) duration of its incidence, or (3) both. In other words,

23.50 tilt of Earth's equator to the plane of the ecliptic and the parallelism of the axis on orbital path of the Earth are major factors of seasons on Earth. There are some important dates which are significant in identifying the periodicity of seasons. This signifies the importance of equinox and solstice on Earth as geometrical relationship with the Sun. Let one assume the beginning of annual cycle of season from March equinox to follow it through the entire period of revolution i.e. one year in context of the Earth. On the basis of this, four dates signifying specific geometrical relationship are important and mark the transition from one season to other. They are 20th or 21st March representing spring equinox, 20th or 21st June representing summer solstice, 22nd or 23rd September representing autumn equinox, and 20th or 21st December representing winter solstice. The Sun Earth geometry on these significant days affecting seasons is discussed below.

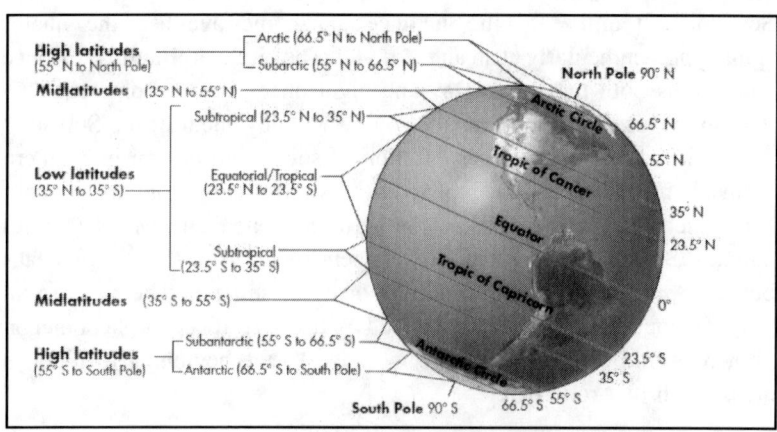

**Fig. 3.6: Important lines of Latitudes and Thermal zones From Arbogast (2014, p. 16)**

**Spring or vernal equinox**[2]: This occurs on 20th or 21st March. At this time the sub perpendicularly making an angle of 900. Solar insolation is most intense **(ref. Fig. 3.5)** there. At this time the circle of illumination touches both the poles of the Earth. Day and night at the time of equinox are equal in duration i.e. 12 hours each. This date also marks the northward movement of the Sub solar point and increasingly longer duration of solar illumination in Northern hemisphere till it reaches tropic of cancer in June. The day of equinox also heralds the beginning of spring season in Northern hemisphere and autumn in southern hemisphere. This may also be said to be the beginning of onset of summer season in areas north of the equator. Days gradually start becoming longer and longer on higher northern latitudes. Correspondingly, nights become shorter. This also heralds dawn (beginning of day) beyond 66.50 north latitude (Arctic Circle) and dusk (beginning of night) beyond 66.50 south latitude (Antarctic Circle). No Sun set takes place in areas north of the Arctic Circle between two equinoxes. It gives rise to a phenomenon known as **midnight Sun** in northern sub polar and polar areas.

**Summer Solstice:** Summer solstice occurs on 20th or 21st June. Around these days sub solar point comes to be located on the **Tropic[3] of cancer** i.e. 23.50 north latitude. In other words, the Northern hemisphere reaches its greatest tilt toward the Sun. It makes the duration of solar illumination longest in areas north of equator.

**Autumn equinox:** On its return journey from the tropic of Cancer (23.50 N. Lat.) the sub solar point revisits the equator around 22nd -23rd September. It implies that the Sun once again comes over head the equator shining perpendicularly at an angle of 900. The circle of illumination once again joins both the Poles. Day and night once again become equal in duration. This date but marks the southward movement of the Sub solar point and increasingly longer duration of solar illumination in southern hemisphere. It heralds the repeat of the similar conditions in southern hemisphere that is experienced in northern hemisphere after March (spring) equinox. Spring season sets in southern hemisphere. Similarly dawn breaks beyond Antarctic Circle marking the onset of summer. The areas south of 66.50 south latitude is characterized by the occurrence of phenomenon known as mid night Sun It conversely marks the beginning of night in areas north of Arctic Circle.

**Winter Solstice:** Post autumn equinox period is characterized by the journey of sub solar point in southern hemisphere till it reaches Tropic of Capricorn on 21st-22nd December. At this time the southern hemisphere has maximum tilt toward the Sun. This is known as Winter Solstice. At this time, southern hemisphere experiences almost identical conditions that northern hemisphere experiences at the time of summer solstice in June.

It is obvious that it is the changing geometrical relationship of the Earth with the Sun that is responsible for seasons and their duration on our planet. There, however, are two classifications of globally accepted seasons - (1) Astronomical seasons and (2) meteorological Seasons.

## Chapter - 3:2:4:3A: Astronomical Seasons

Astronomers and Scientists use The Dates of Equinoxes and Solstices to mark the beginning and end of Seasons in a year. Based on Earth-Sun Geometrical relationship astronomers identify four seasons in a year corresponding with the location of Sub Solar Points:

- Spring the season between March Equinox to June Solstice,
- Summer the season between June Solstice to September Equinox,
- Fall (autumn) the season between September Equinox to December Solstice, and
- Winter - the season between December Solstice to March Equinox.

Astronomical seasons provide a basis to know the positions of different celestial bodies not only within the solar system but also beyond in relation to the changing Earth- Sun geometry. It is found that stars, constellations or even planets are positioned differently in different seasons on the Earth. Knowledge of their exact positions within certain time frame is considered essential for Space expeditions from Earth and associated research. However, changing dates of the equinoxes and solstices has a bearing on the length of astronomical seasons. It can vary within a year as well as between years. It poses a problem in comparing and analysing seasons of different years for regulating activities essential for survival of human beings. Therefore, a different classification based on average weather conditions in different months of the year is generally adopted worldwide. Seasons, thus, classified are known as meteorological seasons.

## Chapter - 3:2:4:3B: Meteorological Seasons

Like astronomical seasons there are four meteorological seasons each having duration of three months but avoiding direct reference to astronomical positions of the Earth vis-a- vis the Sun. Such a classification makes comparisons over any number of years more consistent and facilitates forecasting with better accuracy in order to augment human activities. Duration of meteorological seasons has been defined as follows:

- Spring between 1st March and 31st May,
- Summer between 1st June and 31st August,
- Fall (Autumn) between 1st September and 30th November, and
- Winter between 1st December and 28th February (29th February in Leap Years).

Meteorological seasons, thus, appear to address the transitional weather conditions between equinoxes and solstices in much better way than the astronomical seasons. Both the classifications, however, are better suited for countries beyond the tropics in both the hemisphere due to very nature of configuration of the Earth. The conditions in intra tropical regions due to equatorial bulge in between and apparently faster changes in Sun- Earth relationship there is always a perceptible transition of weather conditions between any two seasons. Therefore, many countries have their own scheme of identification of periods of season, hence the numbers of season. For example, in India six seasons are recognized as Vasant (equivalent of spring season), Hemanta , Grism ( equivalent of summer), varsha (rainy season), Sisir (equivalent of autumn) and Sheet (equivalent to cold or winter season). On an average each of these seasons has duration of two months based on lunar calendar - Vasant (mid Magha to mid Chaitra), Hemanta (mid chaitra to mid mid Jyestha), Grism (mid Jyestha to mid Shrawan), Varsha (mid Shrawan to mid Ashwin), Sisir (mid Ashwin to mid Marg- shirsh) and Sheeth (mid- Margshirsh to mid Magha).

---

### Five Reasons for Seasons

1. Axial tilt of the Earth-If the axis of Earth were not tilted, there would be no seasons.

2. Differential sub solar points-. The sub solar point is the place where the Sun's rays strike Earth most directly. This occurs because the Sun angle is 90° at that point and the rays are perpendicular to the surface.

3. Migration of sub solar points-Due to the axial tilt, the sub solar point migrates between the Tropic of Cancer and Tropic of Capricorn over the course of the year.

4. Occurrence of solstices and equinoxes- solstices occur whenever the sub solar point is at either 23.5° N or S latitude. An equinox, in contrast, occurs when the Sun is directly over the Equator.

5. Changing position of Earth relative to the Sun during its progresses on its orbit.

### Factor Description

❑ Revolution Orbit around the Sun; requires 365.24 days to complete at 107,280 km/ph (66,660 mph).

❑ Rotation (Earth turning on its axis): takes approximately 24 hours to complete.

> ❑ Tilt Axis is aligned at about 23.5°angle from a perpendicular to the plane of the ecliptic (the plane of Earth's orbit).
>
> ❑ Axial parallelism Remains in a fixed alignment, with Polaris directly overhead at the North Pole throughout the year.
>
> ❑ Sphericity appears as an oblate spheroid to the Sun's parallel rays; the geoid seasons are caused by the 23.5° tilt of Earth's equator to the plane of the ecliptic and the parallelism of the axis that is maintained as Earth orbits the sun.

These months of Indian calendar roughly corresponds respectively with mid January to mid March, mid March to mid May, mid May to mid July, mid July to mid September, mid September to mid November and mid November to mid January. The scheme takes into account a sufficiently long periods of transition between seasons generally experienced in tropical and subtropical regions of the world. It takes into account a cycle of feeble positive change to waning change with a maximum of the weather phenomena in between. This is considered to be very significant for agriculture based economies particularly in areas which continue to be dependent on nature.

### Chapter - 3:2: 5: Referencing and Geographic Grid: Latitude and Longitude

As geography is basically concerned with spatial distribution of phenomena on Earth, determination of their location is of prime concern to the geographers. There are many ways in which reference to a location may be made. A very common method is to refer location of a place or phenomenon in relation to some other place or phenomenon e.g. Delhi is located on the boundary of Haryana, Punjab and Uttar Pradesh. Or, Green Land is an island located in the North Atlantic Ocean. Such descriptions of places are termed as **relational locations** and are very general in nature and many times misleading. In scientific studies, however, locations need to be referred with greater precision and international acceptance. Geographers therefore, generally need to have **pinned** or **absolute** locations of places. It is achieved by using geographic grid on the Earth. Due to spherical shape of the Earth the grid has a series of intersecting circles, one set of which extends in east-west direction and the other set from north to south joining the two poles. The sections of grid running in east-west direction are termed as lines of latitude. Similarly, the lines of grid converging at the poles are known as longitude or meridians.

**Geographic grids are, thus, defined as an orderly system of circles—meridians** and parallels—which cut each other at right angles to make a point. They may be used to **locate position on the globe** (Strahler, 2011: 40). All places on the Earth are located on some latitude and longitude. Ptolemy (A.D. 90–168) the geographer, astronomer and mathematician was probably the first person to make the use of the terms latitude and longitude. It was he who divided the Earth in $360^0$ as spheroid. He also divided each degree in subdivisions of 60 minutes and each minute in 60 seconds. It facilitated him in preparation of maps with great accuracy. It, however, needs to be acknowledged that latitudes and longitudes conceptually predated him. It was on this concept only that **Erosthenese** calculated the diameter of the Earth almost correctly.

**Latitudes:** Latitude refers to the **angular distance** of a place measured in relation to any **fixed celestial point**. There are many ways to find out the latitude of a place. The angles made by the **Sun on horizon** in relation to the equator $(0^0)$ give the latitude of that place. This requires adjustments in relation to seasonal variations. Similarly, in northern nights, the angular altitude of **Polaris (North Star)** from horizon gives latitude of the place. In southern hemisphere where North Star cannot be seen help is taken of constellation Southern Cross (*Crux Australis*) in locating a fixed star over the South Pole. Its altitude in relation to southern horizon gives the value of latitude in southern hemisphere. The latitude of a place, however, may easily be calculated considering the centre of the Earth as a fixed point. The equator, representing an imaginary line encircling the Earth represents an angular distance of zero degree $(0^0)$ from the centre of the Earth. It represents the maximum circumference of the Earth and divides the surface of the Earth in two equal halves-

Northern Hemisphere and Southern Hemisphere. It, therefore, is a **great Circle**. The imaginary lines north or south of the equator having angular values of more than zero degree are designated accordingly as north or south latitudes. Away from the equator the angular value of latitudes goes on increasing till the angle becomes perpendicular at $90^0$ representing the latitude for the Poles. Lines connecting all the similar angular values on the surface of the Earth are called **lines of latitude.** Since all the latitudes having equal angular values encircle the Earth parallel to the equator they are also called **parallels of latitude** or simply **parallels**. But their circumference is smaller than that on the equator. Therefore, they are known as **small circles**.

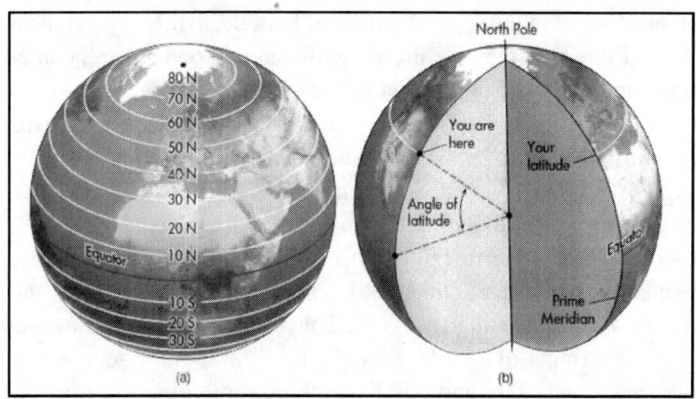

**Fig. 3.7: Lines of Latitude and Latitude (Arbogast: 2014: 16)**

If one takes into account one degree of interval between latitudes including the equator there are 181 lines of latitude (parallels) of which 90 are north of the equator and 90 south of it. Distance between two lines of latitude on the surface of the Earth at $1^0$ intervals is 110 km. (69 miles). Knowledge of the distance between the lines of latitude and its divisions in minutes and seconds gives more specific location of any geographic phenomena. By this method any location may be determined within an error limit of 30.5 metres. There is another way of locating phenomena between the distances of two latitudes more precisely. It is done by converting latitude and its subdivision of minutes and seconds to decimal notations by using a simple equation wherein minutes and seconds are brought to equal numerical value as follows:

Decimal Degrees = Degrees + .d

In which .d = M.m (M/60) + (seconds/60)

For example, if the geometrical extent of India between 80º4'28" and 37º17'53"N latitudes, and 67º7'53" and 97º24'47"E longitudes has to be converted to decimal notations the location will be read as 8.074 and 37.298 N and 68.131 and 97.413E respectively. This system is very frequently used to show the Earth Sun relationship through important lines of latitude as the Tropic of Cancer (23.5° N), the Tropic of Capricorn (23.5° S), the Arctic Circle (66.5° N), and the Antarctic Circle (66.5° S).

In geographical studies, specific latitude and longitude for precise location though are very important; geographers very often tend to study them collectively to discern regional patterns of processes and associated resultants. Clubbing the adjoining lines of latitude, in this context, provides a very useful tool as they are most significant components in geometrical relationship with the Sun and its radiant energy. On the basis of broad characteristics nine geographic zones may easily be identified as (i) equatorial (5° N to 5° S), (ii) tropical (5°-23.5° N and S), (iii) sub tropical (23.5°-35° N and S), (iv) temperate, North (35° - 55° N) (v) temperate, South(350 - 550 S), (vi Subarctic (55°- 66.5° N), (vii) Arctic (66.5°- North Pole), (viii) sub Antarctic(55°- 66.5° S), and (ix) Antarctic (66.5° S- South Pole). These nine regions based on their relative position and similarity of broad physical processes may be clubbed under three groups of latitudes as (1) Low latitudinal zone extending broadly between 35° N and 35° S latitudes combining the tropical and subtropical regions of the Earth. They represent warmest parts of our planet, (2) mid latitudinal zone corresponding with temperate zone of 35° to 55° of latitudes in both the hemisphere. They experience highly variable weather in different seasons of the year, and (3) High latitude zone represented by the area of 55° to 90° latitudes in both half of the globe. They also represent the coldest parts of the Earth.

---

**Latitude: Key Concepts**

1. Lines of latitude run exactly east and west; they are also called parallels because they are always parallel to the Equator and each other.

2. Lines of latitude are determined by measuring the geometric are between two lines projected from the centre of Earth to the surface at the Equator and at the location in question. The reference point for latitude designations is the Equator (or 0°); latitude designations extend to 90° N and S.

3. Lines of latitude never intersect and therefore have the same value over the entire circle.

4. Zones of latitude can be grouped into nine geographic zones. These nine can be further grouped into three broad categories: the low, middle, and high latitudes

---

## Longitudes

Like lines of latitudes longitudes are also imaginary lines. Each of them represents equal angular distances from the centre of the Earth east or west of a selected line on the surface. They run in north south direction converging on North and South poles.

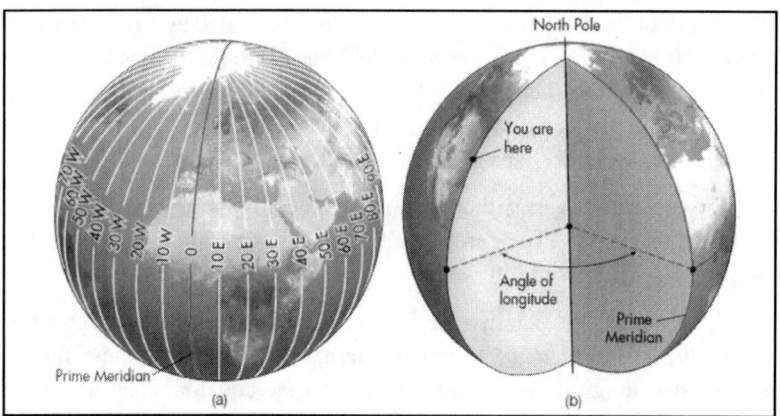

From Arbogast (2014:18)

***Figure - 3.8: Lines of Longitude and Longitude***

Longitudes are also known as **meridians**. The word meridian has been derived from Latin word *meridies*, meaning "midday. The term implies that the time of midday is same on entire length of the line of longitude. It, therefore, may also be said that the term   glongitude is the name of the angle and *meridian* names the line". The line passing through the Old Royal Observatory at Greenwich near London, England was selected to represent **0° longitude** or prime meridian by a treaty in 1884. The lines of longitude east of this prime meridian are referred to as east longitude. Similarly, the meridians west of it are referred as west longitude. If one takes into account an interval of 1° between two meridians, the total number of longitudes is 36°, 179° E. and 179° W. Longitude 0° i.e. the

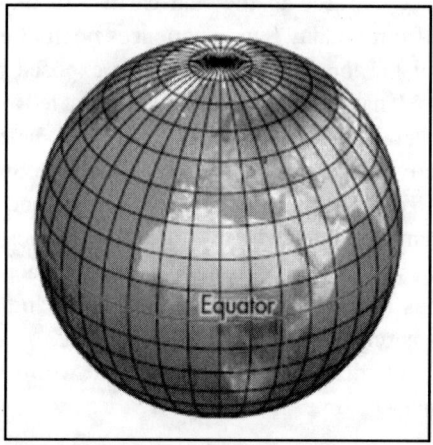

***Fig. 3.9: Geographical Grid.***

prime meridian and 180° are common and are not denoted by any directional notation. It also suggests that each of the lines of longitude has its pair on the other side of the spherical earth forming complete circles of even size and which together can divide the Earth in two equal parts. All of the lines of longitude, therefore, are great circles, like the equator.

These meridians pass through all the lines of latitude and intersect them at right angle. They form a geographical grid. They, however, are not evenly spaced like latitudes.

The distance between two meridians at one degree interval on equator is roughly 111 kilometres (69 miles) whereas at 600 latitude the distance between two longitudes is about half that on the equator. There they are spaced at a distance of 56 kilometres (35 kilometres) only.

## Longitude and Time

There exists a very close relationship between longitude, Earth's *rotation,* and time. The Earth due to its spherical nature has 36° lines of longitude. Its rotation on its axis from west to east, therefore, exposes each of them from east to west one by one to the midday Sun in a cycle of 24 hours. It is this reason that the lines of longitude are also called meridians as mentioned earlier. It implies that in an hour 15 lines of longitude experience incidence of highest solar angle (due to the position of the Sun overhead). This position has universally been accepted to represent 12 noon at that longitude (meridian). This has been the yard stick for local time all over the world. The time on each of the meridians varies by 4 minutes (60°=1 hour ÷ 15). The time is advanced by 4 minutes on immediately eastern meridian as they are exposed to the Sun earlier. In other words, the longitude 15° west from midday Sun experiences noon after one hour. In this manner each of the longitude (meridian) is exposed to the midday Sun once in 24 hours. It has been this understanding that led to establishment of Sun Dials at several places all over the world to measure time and regulates human activities. However, following the high Sun as indicator of time could lead to different time even within a single settlement or area for, its eastern margin would have noon earlier than its western boundary. It required continuous adjustment of time. With the advancing technology, growing interactions and interdependence of places and countries the use of local time was rendered impractical by the end of 18th century. A need was felt to standardize the time all over the world with adequate margin for adjustment of time.

## Time Zones

In view of the above mentioned the International Meridian conference held in 1884 resolved to set 24 standardized time zones roughly of one hour each. Selected standard meridian for each time zone was accepted to represent the standard time for that zone. Because the Earth completes its rotation in 24 hours each hour is equivalent to 150 of longitude (360° ÷ 24 hours = 15°). The conference also resolved to accept the longitude passing through Greenwich to be the Prime meridian (0° longitude) and 7.5° East and West longitude on both sides of it (i.e. 15° of longitudes) to represent its time zone. Similarly, the central meridian for each time zone is fixed. The world time, thus, started being coordinated with what is known as Greenwich Mean Time (GMT). Most of the countries of the world accepted this. Many a time GMT represents official Coordinated Universal Time (UTC[4]). They are also known as Zulu time.

Based, however, on this concept of time a country having greater longitudinal extent may have more than one time zone within their boundaries. Thus, geographical boundary of Russia from west to east extends over ten time zones, Canada and China have four time zones each and the United States of America spans over three time zones. Many

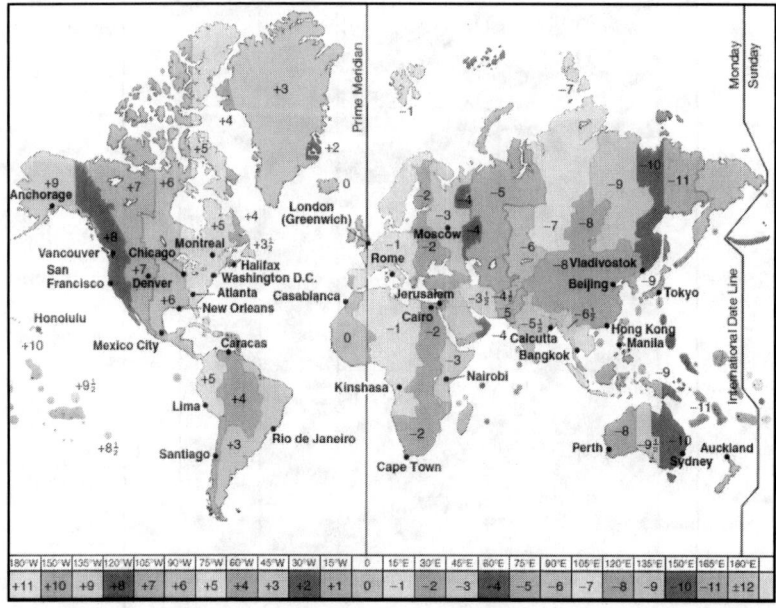

***Fig. 3.10: Time Zones of the World and International Date Line.***

countries like the U.S.A. and Canada adjust the time according to their time zones. Persons travelling in these countries, therefore, subtract one hour if they enter next western time zone.

## International Date Line

As the earth is spherical and perpetually rotating it necessitates assumption of a dividing line from where days are believed to begin. As the Prime Meridian (0°) at Greenwich is accepted to represent the standardized global time its pair 180° meridian on other side of the Earth has a time difference of exactly 12 hours. Therefore, when it is midday (12 noon) at the Prime Meridian it is mid night (12 o'clock) at 180° longitude. As mid night has been accepted to represent the transition between one day and the next day or in ot Fig. 3.11: Positioning International Date Line her words from one date to the other, 180° longitude in general have been selected to represent the International Date Line.

However, the International Date Line has to be substantially modified due to political nature of the area it passes through in the Pacific Ocean. Generally over the water body of the Pacific this line conforms to 180° longitude but it skirts the settled land areas for practical reasons.

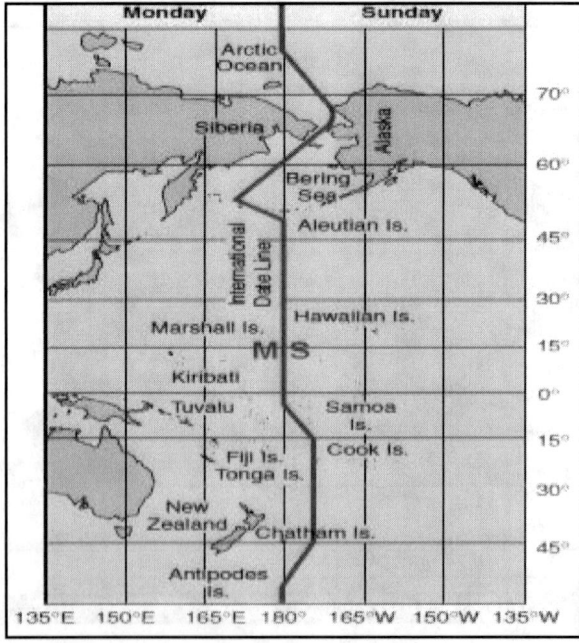

**Fig. 3.11: Positioning International Date Line.**

The benefit of demarcation of the globe on the basis of this line is that any travelling person from west to east has to advance the date by one day to synchronize the time in eastern hemisphere. On the other hand, if somebody travelling east crosses this line has to adjust his date to be one day early. Figure 3.11 shows how International Date Line deviates from the 180° longitude.

---

**LONGITUDE: KEY CONCEPTS**

- Lines of longitude run north and south and are also called meridians.

- Longitude is determined by measuring the geometric arc between two lines projected from the centre of Earth to the surface at the Equator and to the meridian in question. The reference point for longitude is the Prime Meridian (or 0°), which divides Earth into eastern and western halves (eastern and western hemisphere).

- The distance between meridians is greatest at the Equator but decreases steadily until converging at the poles.

- Meridians are located east and west of the Prime Meridian and extend 180° E and 180° W to complete a full circle.

- All lines of longitude converge at the Poles and have its pair exactly at a difference of 180° east or west of the Prime Meridian.

- The pair of longitude together divides the Earth in two equal halves. All the pairs of longitude, therefore, form Great Circles.

---

## References

1. Gabler, Robert E., Petersen, James F., Trapasso, L. Michael, Sack Dorothy (2009): *Physical.*

2. *Geography, Ninth Edition;* Brooks/Cole, Cengage Learning; Belmont, USA.

3. Christopherson, Robert W. (2012): *Geosystems : an introduction to physical geography, 8th ed.*

4. Prentice Hall, New Jersey, USA.

5. Arbogast, Alan F. (2014): Discovering Physical Geography (3rd. Ed.), John Wiley & Sons; NewJersey, USA.

6. Calais, E. Purdue University -EAS DepartmentCivil 3273 ecalais@purdue.edu

## Notes

1. **Christopherson** (2012:14) suggests one stadia to be equal to185 m (607 ft) meaning thereby that **Erosthenese** calculation suggested the polar circumference to be 46,250 km (28,738 mi) which may still be considered to be very close to the presently accepted figure, Following figure is a representation of the Earth's shape and size attributed to the distortion caused by its differential rotational velocity at different latitudes.

2. Name of the equinoxes and solstice has reference to the conditions in northern hemisphere).

3.  The word tropic has been derived from the Greek word tropicus which means to turn or change. (The tropic of cancer or the tropic of Capricorn, thus, represents the line of latitude from which sub solar point turns back to make its return journey to the other tropic)

4.  *In fact, UTC is not based on the concept of meridian. It takes into account the sidereal period of the Earth's slowing rotation.* UTC, thus, represents average time calculation obtained from atomic clocks from all over the world. It, in fact, replaced GMT in 1972. Even GMT is adjusted according to UTC despite the fact that it continues to be accepted as Prime Meridian. Use of GMT for coordinated universal time is now discouraged. or, add one hour if they move to neighbouring eastern time zone. Many of the countries in order to avoid such complications adopt one time based on selected Standard Meridian within their national boundaries. China, for example, has selected one standard time for the whole country despite four hours difference of time between its western and eastern borders. India likewise has chosen 82.50 E. Longitude passing through Nanny near Allahabad as Indian Standard Time (IST) though there is a longitudinal difference of almost 300 between its western and eastern extremes. In addition, the alignment of national boundaries and political expediencies also distort the international time zones even in countries which adopt standard time based on their time zone. It is also affected by the uneven distribution of land and sea.

## Review Questions

1.  How much is Earth's rotation axis tilted away from perpendicular to the planet's plane of revolution around the Sun?

    (a) $0°$           (b) $23\ 1/2°$           (c) $78\ 1/2°$           (d) $90°$

2.  The moment when the Sun is directly over either of the two tropics is called

    (a) Solstice      (b) Equinox           (c) Season           (d) Ellipse

3.  In which month is the Sun directly above the tropic of Cancer ($23\ 1/2°$)

    (a) February   (b) April               (c) June               (d) August

4.  Earth's North Pole is always pointed toward the _____.

5.  The Sun is directly above the equator in the months of _____ and

    _____.

6.  The reason that it is hot in the summer is because the Earth is closest to the Sun. (True or False)

7.  There are 12 hours of daylight at the equator every day of the year. (True or False)

8.  How does Earth's rotation on its axis cause night and day?

9.  With respect to EarthSun relationships, how are tropical latitudes different from middle and high latitudes?

10. For February 1, describe in general terms the latitude zone where the Sun will be directly overhead.

11. What is circle of illumination?

12. Why the shape of the Earth is called geoid?

# Chapter - 4

# Geological Evolution of the Earth

## Objectives

Main objective of this chapter is to trace the attempts that have been made to find out the age of the Earth as well as to acquaint them with distinct periods in its evolutionary history. It is expected to help students to understand the unique character of this living planet in our solar system. It will undertake the study of:

- Attempts to determine the age of the Earth,
- Geological time scale, and
- Major happenings during different geological periods.

## 4:1: History of The Earth

Despite the fact that the Universe is believed to have been formed some 13.7 billion years ago, our solar system including the Earth is suggested to be a very young entrant to the families that occupy the cosmic space. People, since ancient times have been inquisitive about the mode of formation (*discussed in Chapter-2*) and the age of the solar system and its constituents. Many believed that the Sun might be the parent body of relatively recent origin from which planets including the Earth were formed. Due to lack of adequate information and technology it were generally conjectures about the 'creation' based generally on theological teachings that scholars attempted to place the age of the Earth. It was in this vein that **Archbishop Usher** of Ireland on the basis of genealogy as given in Bible concluded in 1654 that the *Earth* was hardly 5500 years old. He believed that it was created on October 26, 4004 BC precisely at 9:00 am. There were very few takers of such a short life span of the Earth even then. Advances in the field of knowledge and technology in post Exploration and Renaissance periods equipped the scholars to have more rational methods to view the origin of the Earth and its features. **James Hutton** (1785) basing his ideas on his famous **'Principle of Uniformitarianism'** though believed the Earth to be very old could not provide a basis for articulation about the age of the Earth. Yet many scientists in the light of Hutton's 'Principle' found ways to estimate the age of the Earth. They adopted different methods for measurement. One of the earlier studies in this respect by **John Phillips** (late 18th century) was based on the measurement of rate of sedimentation, thickness of sedimentary strata and age of fossils obtained from them. His calculations placed the age of the Earth around 96 million years. **Lord Kelvin** (1862), a physicist, assuming that the Earth was initially in a molten state invoked the rate of cooling of the Earth to ascertain its age. Using thermal gradient on the surface of the Earth he estimated its age to be about 100 million years. His conclusion, however, was rejected by 1895 on the basis of not considering the fluid and viscous mantle underneath the Earth's crust and its impact on thermal gradient on the surface. **John Perry** acknowledging the impact of convective motions of the mantle under a thin crust estimated the age of the Earth to be between 2 to 3 billion years in 1895. However, Kelvin did not concur with such a methodology. Instead he revised his estimate in 1897 reducing the age of the Earth between 20 and 40 million years. **George H. Darwin**, son of **Charles Darwin** and an astronomer based his calculation on the tidal friction that was generated after supposed separation of the Earth and Moon in a molten stage and which gave the Earth its 24 hours day. He calculated the age of the Earth to be 56 million years. He supported the method of calculation adopted by Kelvin. **John Joly**, on the other hand, attempted in 1899-1900 to determine the age of the Earth by establishing the rate of salinity increase in the oceans assuming that the oceans and continents were formed simultaneously. He came to the conclusion that the Earth could be around 90 million years old. Till the early parts of 20th century, thus, most of the scholars placed the age of the Earth ranging between 3 million years and 150 million years only. However, many of the contemporary scholars like the geologist **Charles Lyell** (1872) were critical of such short estimations of the Earth's age. They held the view that the Earth had been undergoing tremendous geomorphic and climatic changes all through its geological history. And though the process as suggested under the principle of uniformitarianism continued to form the basis of explanations of the characteristic features on the Earth's surface it came to be realized that they, consequent upon **eustatic** and climatic changes, underwent substantial changes in intensity of operation throughout the geological history of the Earth. It increasingly became evident that the climatic changes would impinge on the rate of sedimentation as it would influence fluvial processes as evidenced in many areas since hydrological cycle was integrated in the Earth system. Erstwhile humid areas are found to have become arid or semi arid or extensive glaciated areas during certain epochs of the Earth history have been undergoing relocation. Similarly, the processes of diastrophism and volcanism have been found to have been more active spatially as well as temporally during some periods in the course of the Earth's evolution (*see continental drift and Plate Tectonics in chapter IX*) . Therefore, basing estimates on a single variable like that of the rate of sedimentation, the rate of cooling or salinization would not give a valid result for the age of the Earth. Even for biologists who believed in **Charles Darwin's** estimation in 1859 of earth's age to be 100 million years or so on the basis of **faunal succession** considered it to be too small a period to affect random heritable variation and process of selection inherent in his **Theory of Evolution**. Keeping in view the findings of studies in modern biology it would take about 3.5 to 3.8 billion years for earth organisms to evolve to their present stage. Therefore, the age of the Earth could not be less than that.

By the end of 19th century, discovery of radioactive elements in earth rocks by husband wife duo of **Marie** and **Pierre Curie** (1898) brought forth the fact that the Earth was capable of replenishing heat due to decay of radioactive elements. It requires to be noted that **radioactive elements** are unstable and decay at a fixed and calculable rate till ,they stabilise as nonradioactive matter. Radioactive elements, however, do not decay directly to stable nonradioactive matters. They pass through a series of decay during which hidden heat of the original element is released. The knowledge underlined the fact that dissipation of Earth heat and its associated physical phenomena as invoked by many scholars earlier could not provide the basis for measuring the age of the Earth. It was **Rutherford** and **Soddy** who discovered in 1904 this nature of radioactive elements. Referring to the period that half of the original radioactive element takes to decay to other element they coined the term **'half life'**. Some of the radioactive elements like *uranium* and *thorium* decaying to *lead* have been found to have a very long half life (*Uranium 238 to Lead about 4.5 billion years*). Similarly, half life of decay of potassium to argon is calculated to be about 1.3 billion years. Most of the elements with short half lives seem to have decayed into stable matters found on the Earth. It, thus, was believed that by finding the rate of decay of one radioactive element to other series could give the age of the Earth. This process is called eradiometric dating'. **Rutherford** in association with **Boltwood** after studying the decay series of 26 rock samples in 1905 suggested the age of the Earth to be somewhere between 92 to 570 million years. It was later found by **Boltwood** that the technique adopted for such estimation was faulty. A more refined technique adopted for the same 26 samples suggested that the Earth could be 2.2 billion years old. Dissatisfied with the basic assumptions of **Rutherford** and **Boltwood** that radioactive decay was trapped in rocks as helium atoms **Arthur Holmes**, a geologist, adopted *uranium- lead* series dating. He persisted with his experiments till 1931 when as a member of the **National Research Council** of the US he suggested the age of the Earth to be about 3 billion years.

Since 1960s the technique of radiometric dating has been more refined. This was also realized that the crust samples of the Earth 'could be contaminated due to exposure to various processes that characterize the geo-system. In order to avoid the problems arising out of perceived contamination in Earth rocks and its impact on the processes of determination of age of the Earth **Patterson** (1956) studied **uranium-lead** and **lead-lead** isotopes in **meteorites**. He came to the conclusion that the Earth could be as old as 4.55 billion years. It may be of interest to know that the oldest rocks found in Western Australia and Canada is dated to be between 4.2 and 4.4 mya. The estimate is very close to presently accepted age. Recent dating efforts taking into consideration **plate tectonics, weathering** and **hydrothermal circulation** as well as the oldest rock samples establishes the age of the Earth to be about 4.6 billion years with almost negligible margin of error.

### 4:2: Geological Time Scale

The **Geological Time Scale (GTS)** refers to chronology of evolution of rock strata. It may suggest two kinds of time: **relative** and **absolute**. Relative geologic time scale reflects the sequence of events in the wake of rock formation and their placement over each other. It, thus, reveals the relationship of events in relation to the formation of rock strata from older to the younger ones and what has been occurring throughout the life history of our planet. It is based on the **principle of superposition** of undisturbed rock strata in which younger rock strata are always found overlying the older ones. It forms the basis of study of what geologists call **stratigraphy**.

On the other hand, absolute time scale denotes the actual number of years in relation to the present and which scientists keep on improving

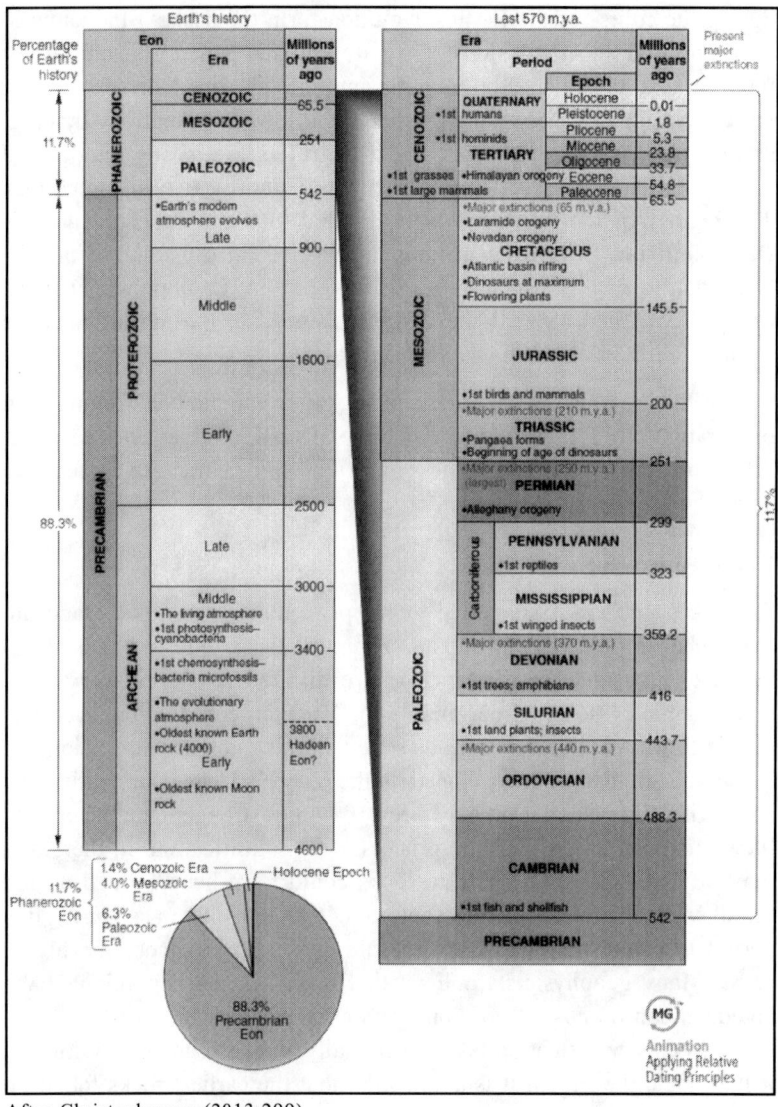

After Christopherson (2013:299)

*Fig. 4.1: Geological Time Scale showing highlights of Earth's history*

with adoption of more efficient methodology. It helps in establishment of relative time scale. Due to vastness of time believed to cover about 4.6 billion years presently the geological time scale is organized into various temporal units and their subdivisions. Major chronostratigraphic (position) and geochronologic (time) events that are believed to have taken place during the course of the Earth's evolution form the basis of the name of each unit and its subdivisions. They are unitized generally from largest units to smaller units within one geochronologic (time) division. Thus, largest division has been named supereon followed sequentially by eons, eras, periods, epochs and ages. Supereon refers to a vastly long period of time in the history of the Earth. It may comprise two or more eons. The history of the Earth has been divided into one supereon namely **Pre Cambrian**. No sufficient time has elapsed since the end of the Pre Cambrian to qualify as supereon. In fact, the Pre Cambrian is believed to have occupied 88% of the Earth's history i.e. the period since its formation till 542 million years ago (mya).

Following chart, arranged sequentially after International Commission on Stratigraphy (2009) and Christopherson (2013), gives an idea about different units of GTS and their duration with characteristic features since the origin of the Earth.

## Pre Cambrian Super Eon

The Pre Cambrian is suggested to comprise three eons of (a) Haddean, (b) Archaean, and (c) Protorozoic. Amongst these only two later eons-the Archaean and the Protorozoic are divided into eras. Very little stratigraphic evidences are available for Haddean eon. It is believed that the Earth was mostly in a gaseous state with much bigger dimension and it magnetic field had not developed till the early Haddean. During this time solar wind flux which must have been 100 times stronger than the present blew most part of the Earth in process of its solidification. It, therefore, may be assumed that very little rocks could have been formed during initial solidification of the Earth between 4600 and 4000 mya. During this period, therefore, there had to be very thin and scattered patches of evolving crust. Many geophysicists believe that it was in the beginning of the Haddean eon (almost 30 million years after the formation of the Earth) that a Mars size meteor collided tangentially and detached one portion of it that made the Moon. It is noteworthy that the earliest rocks found on the Earth is believed to have been formed most likely during Haddean era some 4280 mya.

The Archaeans are believed to have four eras namely (i) eoarchaean, (ii) paleoarchaean, (iii) meso archaean and (iv) neoarchaean. Most scientists believe that only by the end of the eoarchaean i.e. about 3.5 mya, magnetic field was created around the Earth. Formation of magnetic field restricted the escape of gaseous molecules from solidifying Earth.

As they were arrested in the gravitational field of the Earth it led to the formation of early atmosphere. Similarly, the protorozoic eon has been divided into three eras. They are named as (i) alaeoprotorozoic, (ii) mesoprotorozoic, and (iii) neoprotorozoic. It is most likely that solidified crusts configured as **Super Continent** for the first time by neoprotorozoic era some 700 mya, There has been a cycle of break ups of continents and their reunion since then.

These eras, however, are generally not in common use. From Haddean to Protorozoic they are collectively named as Precambrian. Most of the oldest found Precambrian rocks of the Earth belong to the Archeans. They are characterized mostly by the presence of crystalline rocks of igneous origin and metamorphism of both igneous and sedimentary rocks. They are dominated by gneiss and schist rocks. Supposedly the entire core areas ( **craton**[1] or continental shields) of existing land masses the Peninsular India, Brazilian and Canadian shields, Scandinavian highlands, Chinese and Russian Massifs as well as the parts of Western Australia and Antarctica (which form the nucleus of present continents) have dominance of exposed Precambrian rocks of both igneous and sedimentary parentage (Fig.4.2 ). They are generally distinguished from later formations either by faults or fault lines or mobile belts. In India, igneous archaean rocks are mostly found in present Tamil Nadu, Andhra Pradesh, Chattisgarh, Jharkhand and Rajasthan.

Dharwar system of rocks in India also belongs to this group of Archeans. Intensely metamorphosed rocks of sedimentary origin but undistinguishable from the underlying gneiss and schists of the time are generally found in India in the present states of Karnatka, Madhya Pradesh, Jharkhand, Meghalaya and Rajasthan. They like their predecessor also have remained more or less unaffected by the later earth movements. The crust of the Earth during this phase of Earth's history is believed to be irregular patches and much thinner than the present. This phase is believed to have covered almost the first 4000 million years to 2500 million years in the evident geological history of the Earth.

Investigations of rocks reveal that no form of life existed during archaean eon. It is, however, believed that atmosphere which started being formed some 3500 mya consolidated during later parts of the pre Cambrians during protorozoic eon covering latter 2500 million years. First of the orogenic movements that probably created the Aravallis over archaean basement of Rajasthan-Bundelkhand craton of India is believed to have taken place during early protorozoic eon (Sharma, 2009). Vindyan and cuddappa system of rocks in India were most likely formed towards the second half of the Protorozoic eon. There are also evidences that the Vindyan and Cuddapah basins over the existing underlying crystalline rocks underwent sedimentation with calcareous and arenaceous deposits in closing quarter of the protorozoic eon. They were ultimately uplifted during two phases.

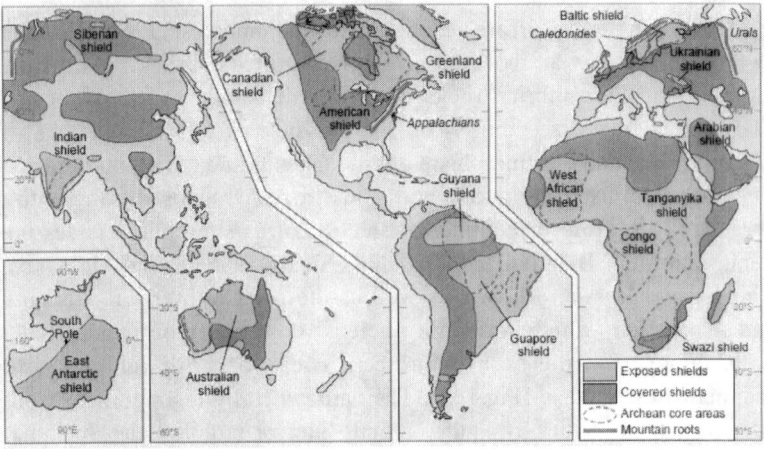

Adopted from **Strahler. A.N.** (2011:387)

*Fig. 4.2: Continental shields*

During Pre Cambrian the patch and thin crustal layers, it is believed, united with each other some 700 mya to form the first of the Super Continents named **Rodinia**. Since then there are evidences that continental land masses have broken at least two times and reunited with different configurations governed by endogenic forces to form super continents. Disintegration of Rodinia is believed to have led to creation of Super **Panthalssic** Ocean. Fragmentation of Rodinia also led to collision of dismembered land masses and formation of mountains. Subsequently the land masses reassembled some 600 mya. for a short period in polar region as another Super Continent named **Pannotia** (see figure 4.3). Pannotia also disintegrated and its three major fragments drifted towards north. By early Palaeozoic era of Phanorozic eon they formed what is named as **Laurentia** (North America), **Baltica** (northern Europe), and **Siberia**. Process of arrangement of continental masses, it is suggested, is continuous and cyclic. It is well established by Geologic Time Scale as shown in fig. 4.1

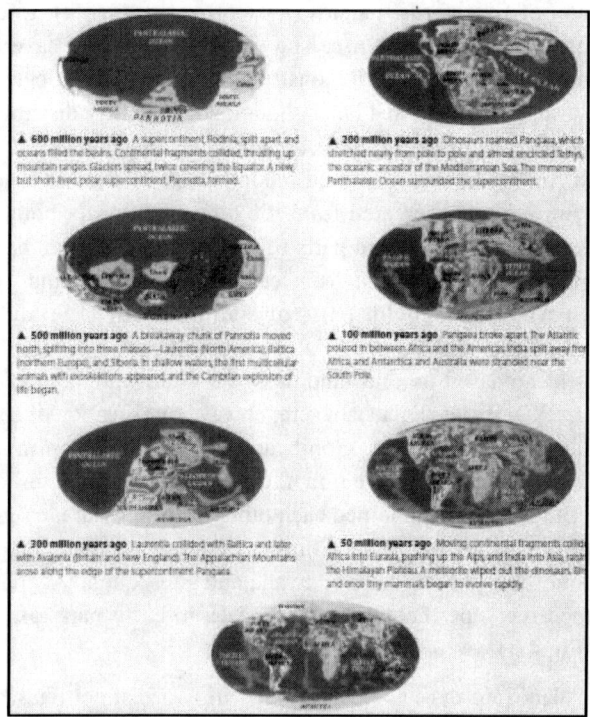

Adopted from Strahler (2011:400)

*Fig. 4.3: Continents through geological Times*

## Phanerozoic or Cambrian Eon

The end of the Pre Cambrian is suggested to have ushered in the present eon named Phanerozoic. This current eon has been subdivided into four most accepted nomenclatures of eras. They are generally named on the basis of life forms for which sufficient geological evidences are available. The four eras have been named as (i) Palaeozoic, (ii) Mesozoic, (iii) Cenozoic, and, (iv) Quaternary. Each of these eras has further been divided into periods[2] based on stages of change in life forms. These periods have further been subdivided into epochs marked by some characteristic formations or deformations on the surface of the Earth including the six major extinctions of life forms during the present eon i.e. in last 542 million years.

## Palaeozoic Era

First era of the contemporary Phranozoic eon has been named the Palaeozoic or the Primary era. The term has been derived from the combination of two words- palaios (ancient) and zoe (life). Literally the word means early or ancient life. The era is estimated to have extended from about 542 to 251 mya. Reconstruction of the land blocks in early Palaeozoic suggests that most lands though not regularly distributed were then in southern hemisphere. The cores of Scandenavian and Irish shield and North American shield formed a continuous land mass whereas the ancestral Europe was separated from it by a narrow water body. Most of Siberian massif, Mongolia and parts of western China were then located in tropical latitudes separated from each other by contemporary seas. The South American shield, parts of south eastern U.S. and Central America, Arabian, Indian, west Australian and Antarctica shields then were mostly amassed as one landmass the Gondwanaland (**Burchfiel** et.al. 1982: 237). By late Palaeozoic era, however, unification of continental blocks started to take place. Gondwana land though continued to be around the southern polar region the northern landmasses of North America, Europe and Asia joined each other to form what later geologists termed as Laurasia (ibid:238). Coming of these northern land masses together and their joining the Gondwanaland subsequently is believed to have formed the super continent of Pangaea in early parts of Mesozoic era (**see Fig.4.3**)

The Palaeozoic era has been divided in two distinct phases namely (1) Lower Palaeozoic, and (2) Upper Palaeozoic having six periods. Lower Palaeozoic is believed to comprise of three periods[3] named as (i) Cambrian, (ii) Ordovician, and (iii) Silurian. The Upper Palaeozoic consists of (iv) Devonian, (v) Carboniferous and (vi) Permian. Evidences from lower Palaeozoic to the upper reveal evolution of life forms from simple to growing complexity as well as gradual spread of predominantly marine life forms to terrestrial environment (**see table: 4.1**). However, the era is also characterized by three major periods of extinctions of life forms during Ordovician, Devonian and Permian periods. Such episodes of extinction may be attributed to changes in ecological conditions caused either by the mobility of the continental blocks inducing orogenic and volcanic incidences, change in energy balance of the Earth and variability in oceanic levels (marine transgression and regression). Two of the major mountain building periods are believed to have taken place during the Palaeozoic era. The Caledonian mountain building episode (almost contemporary to the Vindhyans in India) in early Palaeaozic gave rise to the present highlands of North Ireland, Scotland and Scandenavia. In India it is represented by the mountain system running from east to west from Kaimur hills of Bihar- U.P. to Satpuras in Madhya Pradesh- Gujarat.

Another of the mountain building period i.e. Hercynian mountain building episode took place during the Permian period in which were formed the Applachians of North America, Altaid mountains, Central Massif of France, the Vosges, Black Forest and the Ural mountains of Russia in Europe and most mountains of Asia north of the Himalayas (Tien Shan, Kyunlun etc.)

The era particularly during its early part experienced large scale marine transgression leading to rich formation of limestone. Similarly, Devonian to Permian periods is characterised by the presence of rich deposits of coal in many parts of Gondwanaland including India. It is also during this era that salt deposits came into existence due to prevailing hot climatic conditions in Laurasian parts of North America and Europe.

## Mesozoic Era

Mesozoic era refers to the secondary stages in the evolutionary history of living organisms. Literally it would mean medieval life (mesos= middle+ zoe= life). The era is believed to extend from 251 to 65.5 mya with three distinct periods in between. In early Mesozoic the super continent Pangaea was formed with evolving Tethys Sea in between the northern and southern land masses. Eventually the two land masses started being separated and fragmented by the middle of the Mesozoic.

Periods that represent this era are Triassic, Jurrasic and cretaceous. The era is characterized by the evolution of mammoth reptiles (Dinosaurs) and oversized birds in post Triassic Periods. It is also during this era when the super continent of Pangaea attained its conceptualized form most likely in Triassic period between 251 and 200 mya (Christopherson; 2013). The period towards its end around 210 mya is also believed to have experienced a major extinction of life forms. By the beginning of Jurrassic period first species of birds and mammals evolved. During following Cretaceous period, widespread and thick coniferous and broad leaved trees dominated the flora in the world. This floral capital supported the oversized dinosaurs and birds that dominated the fauna of the time. Most of the land and sea animals, however, perished by the end of the Cretaceous period. The cause of this extinction though continues to be a mystery many scientists believe that the change in atmospheric composition during the period might have been the cause behind the extinction of mammoth animals of reptile family.

This period is also attributed to have immense sedimentation in geosynclines to propel the relative movement of the Gondwana and the

Laurasian blocks and gradual elimination of the Tethys Sea in between. The Jurassic epoch is also attributed to have experienced rifting of the Laurasia causing widening of the Atlantic and separation of the North American block from Eurasian block which culminated during the Cenozoic era. This is also suggested to have resulted in the Nevadan and Larmide orogeny at the end of the Cretaceous in Western North America. Associated volcanism might have been instrumental in changing the atmospheric composition during the transition between the Mesozoic and Cenozoic eras. It was during this time that the Deccan Trap of India which covers an area of almost 500,000 sq. km. with variable thickness in Maharashtra, Gujarat, Madhya Pradesh and Andhra Pradesh was formed as a result of large scale volcanic fissure eruption. Contemporary lava deposits are also found in parts South Africa.

The era is characterized by rich deposits of minerals in different parts of the world. Many of the iron ore deposits in Europe owe their origin to stratigraphical changes brought during Jurassic and Cretaceous periods. A rich deposit of coal, petroleum and gas is found in North American cretaceous rocks. Volcanic dykes created during Jurassic and Cretaceous earth movements enriched Africa with diamond deposits.

**Cenozoic Era**

Most recent of the eras is the Cenozoic (also spelled as Cainozoic, Caenozoic, Kainozoic) . Cenozoic literally means 'new life'. It is dated to have heralded the newest of the era some 65 mya. During this era Mesozoic mega organisms of reptile family were replaced gradually by mammals. More and more areas started being covered by deciduous trees, flowering plants and grasslands. Mammals like horses, cows, elephants, dogs etc. evolved during this time. Recent discoveries reveal that the first hominids (human family) the predecessor of present day human being 'the Australopithecus' evolved during this era some 3.9 mya. Subsequently, Homo habilis (2.8 mya), Homo erectus (1.8 mya), Homo heidelbergenesis (0.6 mya) and the latest the Homo sapiens (0.2 mya) evolved to occupy the global space.

The Cenozoic is normally divided in two periods - Tertiary (between 65 and 2.6 mya) and Quaternary (2.6 mya till date). It is also believed that over 90 percent of the landforms developed during this era.

The Tertiary period has been divided in five distinct epochs of Palaeocene, Eocene, Oligocene, Miocene, and Pliocene. It may be noted that all the epochs under Cenozoic era has a suffix cene. It has been derived from Greek word that means 'recent'. Beginning of the Palaeocene (ancient recent) is characterized by the continuation of events of late Cretaceous with well defined rock strata. Parts of Pangaea- Gondwanaland and Laurasian continents which had started to disintegrate during carboniferous-Permian periods continued to break apart. Continents started to assume their present positions. Indian and Atlantic Oceans widened. As a result African block moved towards southern Eurasia. At the same time, the North American block broke away from the Eurasian block and came closer to eastern Siberian region. Westward movement of the American blocks also initiated the formations of the Andes in South America and upheaval of cordillera of western North America along the continent-ocean margins. Almost simultaneously, eastern blocks of Gondwanaland that included Turkey, Persia, Peninsular India and most likely China drifted towards Eurasia gradually closing the expanses of the Tethys Sea. Collision of Gondwana blocks with Eurasian block (continent-continent margins) begat the formation of highest mountains that include the Alps and Carpathians in Europe, the Himalayas in Asia, and the Atlas in north western Africa since the Eocene epoch. Northward movement of Australia and its collision with the fragments of the Asian plate produced mountains of eastern Asia and Indonesia. All these events are dated to have taken place between 54.8 and 1.8 million years covering Eocene (dawn recent), Oligocene (little recent), Miocene (less recent) and Pliocene (more recent) epochs of the Tertiary period (**Dey**, 1968:113). Peak of the Tertiary mountain building was attained during mid-Miocene. The orogenetic thrust gradually weakened thereafter and came to a close in India with the formation of the Siwaliks and its outliers by Pliestocene epoch of early Quaternary. This mountain building period like any other earlier ones is also characterized by massive intrusion of magmatic materials and displacement of earlier rock strata. Thus, one finds extensive layers of lower Palaeozoic to mid- Mesozoic rocks in parts of new fold mountains along with sedimentary rocks of Tertiary origin in their periphery. A number of minerals like petroleum, and gas as well as legnite are found in Tertiary rocks and where ever igneous intrusion during the mountain building period took place one finds deposits of tin, copper and silver as in case of Bolivian Andese.

Fossils obtained from the rock strata of the period are found to suggest mass extinction of land and water bound mammoth animals of reptile family. The extinction episode was probably induced by a thermal maximum due to extreme changes in carbon cycle which over took the world during the transition between the Mesozoic and the Cenozoic for a brief period of about 2 million years. However, it is during this era that the mammals started to dominate the land life. It is also believed that the first hominide appeared during this era some 3.9 mya. The era may also be called the age of flowering plants as they increasingly spread over land.

Newest of the Cenozoic era is the Quaternary period. Quaternary period has been divided in two epochs of Pliestocene and the Present (Holocene). The Pliestocene which is believed to have its beginning some 2.6 mya, ended some 10,000 years ago. The epoch is characterized by extensive glaciations which covered most parts of the northern land mass and small parts of southern continents. Four ages of glaciations namely Gunj, Mundel, Riss and wurm interspersed the Pliestocene. The epoch, therefore, is also known as glacial period. Last of the glaciations most likely ended between 12,000 and 10,000 years from the present.

End of the glacial episode heralded the present epoch- the Holocene. It is believed that modern man developed during this epoch. Many scholars like to rename it as Antropocene. However, there are scholars who like to divide the Holocene in two parts based on the level of human development and his ability to manipulate the nature. Therefore, some like the introduction of agriculture to be recognised as the beginning of Anthropocene whereas there are people who like to name Post Industrial Revolution Period to be the beginning of Anthropocene.

## Review Questions

1.  What are the units of the geologic time scale? Why is the Cambrian period important?
2.  Give a concise account of the Standard Geological Time Scale.
3.  Give the probable age of the Earth and the methods by which this has been determined.What were the erroneous assumptions upon which were based the three major eighteenth century attempts to estimate geologic time?
4.  To what era of geologic time does each of the following period belong?
5.  Permian
6.  Tertiary
7.  Cambrian
8.  Jurassic
9.  Devonian
10. Cretaceous
11. Name the segments of geologic time in which amphibians, the reptiles and the mammals dominated the land life.

12. What are shields and cratons? List the major ones. Where do the oldest rocks in India occur?

13. When and where in the geologic past did continental glaciations occur?

14. Evaluate permanency of continents and ocean basins since Haddean eon.

15. Trace the periods of glaciations on the Earth.

16. Give an account of mountain building periods with reference to Geological Time Scale.

## References

1. Burchfiel, B.Clark, Robert J. Foster, Edward A. Keller, Wilton N. Melhorn, Douglas G. Brookins, Leigh W. Mintz, Harold V. Thurman (1982): Physical Geology; Charles E. Merrill Publishing Co.; London.

2. Christopherson, Robert W. (2013): Elemental Geosystem (7th ed.), Pearson Education, Boston.

3. Dey, A.K. (1968): Geology of India; National Book Trust, India; New Delhi.

4. Sharma, R.S. (2009): *Cratons and Fold Belts of India*, Lecture Notes in Earth Sciences 127, DOI 10.1007/978-3-642-01459-8_2, C_Springer-Verlag Berlin Heidelberg 2009.

5. Strahler, A.N. (2011): Introducing Physical Geography (5th edition); John Wiley & Sons; New Jersey.

## Notes

1. Cratons are the continental crust that attained tectonic stability after their formation of Cambrian orogeny. In later part of the protorozoic i.e. neoprotorozoic era there are evidences testifying to at least four periods of worldwide glaciations. These periods of low to very low global temperature may be attributed to the changing heat balance of the Earth consequent upon the formation of magnetosphere on one hand and crystallization processes of earth materials on the other. It is substantiated by the fact that most of the ferrous ores of iron, manganese, uranium, chromium, lead, zinc and gold along with mica were formed in the Pre Cambrian rocks due to changing patterns of mineral contents in process of crystallization and metamorphism of matter since Eoarchean era. Formation of atmosphere and associated hydrosphere, it is believed, begat the first organisms in the form of cyanobacteria as well as the first plants of algae family by late protorozoic.

2. Periods have been generally named after the places (mostly from Europe) from where the first representative fossils were found.

# Chapter - 5

# Earth Structure and Earth Materials

## Objectives

Main objective of this chapter is to make students understand the basic relationship that exists between composition of the Earth and its surface features and landscape. The chapter, therefore, is devoted to the study of Evidences about the structure of the Earth.

## 5.1: Evidences About the Structure of the Earth

There are though sufficient information about the matters that compose the upper parts of our planet, knowledge about interior composition of the Earth having a radius of 6371 km. is mostly based on indirect evidences obtained from the study of meteorites, intensity of the movement of seismic waves, gravity anomalies, magnetic polarity in rocks, and variable temperature conditions.

**Composition of meteorites and the Earth:** Analysis of meteorites and meteoritic dust provides an idea about the matters that are found in our cosmos and which are also considered to be the building materials of stellar as well as planetary bodies. Residue of many meteorites reveals that their nucleus has predominantly been composed of heavy matters. In other words, it is believed that the initial formation represented by the inner structure of the Earth must have been dominated by primordial heavy matters with abundance of nickel and iron. The belief is supplemented by the findings of palaeo magnetic studies.

**Seismological evidences:** Evidences obtained from the analysis of the movement and velocity of earthquake waves though indirect, have made significant contribution towards understanding of the Earth's composition and its internal structure. Two of the waves generated by the incidence of Earthquakes **(for details see the chapter on earthquake)**

are of utmost significance in understanding the internal structure of our planet. These waves are named (i) condensational (=compressional or longitudinal or P-waves) and (ii) distortional (= shear or transverse or S-waves) waves. These two waves travel at different velocity. P-waves moving in the direction of propagation are faster than the transverse Swaves. These two waves become very distinct on a seismograph when the recording station is more than 1000 km. from the **epicentre** (the point perpendicular to **focus** or **hypocenter** from where earthquake is generated). Any seismograph station would record the P-waves first followed after some time by the S-waves even if they travel trough a homogeneous body. Knowledge of velocity and difference of time in arrival of the two waves enables one to locate the epicentre of the earthquake. A third type of waves more regular and of high amplitude may also be identified on the surface around the epicentre. They are known as surface waves (= long or L-waves). Velocity of these waves is very closely related with the composition (density) and thermal properties of rocks that make our Earth. It is found that these waves move with greater velocity when they are transmitted through cooler and rigid masses. Conversely, their velocity reduces proportionally when they pass through hotter, less dense and molten segments of the Earth. It is estimated that velocity of these waves increases from about 6-7km/second from focus to a little over 8 km/second towards epicentre (a cooler point relative to the focus). The velocity also increases with depth as earthquake waves encounter increasingly solid and denser materials till the core of the Earth is reached. However, their velocity is found to be decelerating when they pass through more plastic and molten sections just below the Earth's thin crust known as asthenosphere. It is also referred to as 'low velocity zone' in relation to earthquake waves. Similarly, these waves slow down when they pass through the outer core. As the S-waves which can be registered only in solid bodies gradually become weak and are completely lost at a depth of about 2900 km the boundary between mantle and outer core of the Earth. It is therefore, believed that the outer core may be composed of liquefied heavy matters.

The P-waves, on the other hand, are capable of passing through all the phases of matters though with variable velocity. Therefore, they can be recorded even on stations antipodal to the epicentre. However, when these waves come across a boundary between different types of rock and different densities only part of their energy passes through the boundary. In the process, they are refracted in the manner in which light is refracted when it passes through mediums with different densities. Angularity of refraction, however, depends on the ratio of velocity of rock types which is linked with their density and plastic properties (**see figure 5.1**). It results in a part of the Earth registering no disturbance on account of earthquake. In fact, no seismic vibration is recorded between 102° and 143° of the epicentre due to changing directions of refractive waves. It is known as earthquake shadow zone. Rest of the energy is reflected backwards from the boundary of two rock types. Analyses of the velocity and deviations in the direction of earthquake wave propagation (seismic tomography) have enabled the scientists to deduce the structure of the Earth. Following diagram is suggestive of internal structure of the Earth on the basis of behaviour of the seismic waves.

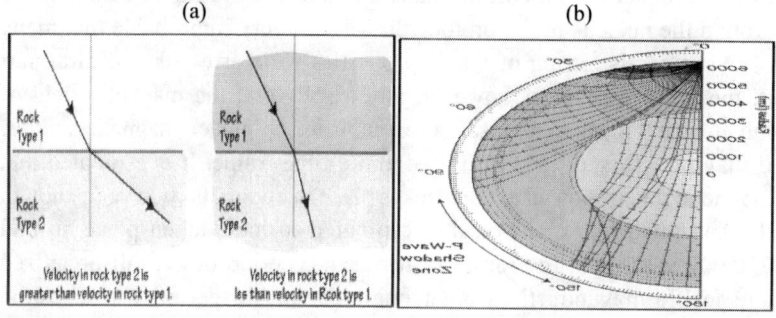

(a) Refraction in two types of rock layers

(b) Generalized movement of seismic waves within the Earth

*Fig. 5.1: Movement of Earthquake waves*

**Gravity anomaly and Earth's internal structure:** Geophysicists had been investigating that how earth features could remain stable on a rotating Earth? In other words, if first order features (continents and ocean basins) and second order (mountain, plateaus plains) had same composition on an eternally moving Earth they would have collapsed. No such collapse has been recorded in the history of our planet so far. On the other hand, **Pierre Bouguer** during his survey in the Andes in 1735 realized that the size of earth features do not necessarily reflect their mass (quantity reflected in weight i.e. pull exerted by the Earth's gravity). This gravity anomaly popularly known as **Bouguer anomaly**, has generally been found to be negative over topographically high mountains and increasingly positive below sea level. It implies that the continent making rocks are having mass deficiency (reflected by their lower density) below them in comparison to the rocks below sea and ocean floors. Negative anomaly is proportional to quantity of displaced massive rocks below them. On the other hand, very less or no matter is displaced below the ocean basin. Now it is well established that different crustal features maintain floatation equilibrium on a denser lower strata (for detail see module on Isostasy).

**Magnetism and Earth's internal structure:** Scientists have been attempting to understand the mechanism that makes electromagnetic field around the Earth functional. Now it is well established after **Elsasser**, a geophysicist who in 1958, suggested that the magnetic field of the Earth was controlled by its liquid outer core. It is important to remember that when the electromagnetic field was created some 3500 mya, the Earth was still in its formative stage of solidification. Before the creation of the earth's magnetic field cosmic matters could be accumulated unimpeded around the nucleus of the protoearth. It is not very improbable that many of the minerals having magnetic properties were arrested in central part of the Earth then. It is, however, noteworthy that the magnetic field of the Earth is subject to reversals as exemplified by palaeo magnetic studies. It may reverse suddenly with no definite time frame. It is estimated that magnetic reversal on an average takes place in about 200,000 years though the last incidence of reversal is estimated to have taken place around 730,000 years ago. In process of reversal, a period of very disorganized and feeble magnetic field is encountered. It is a period when cosmic radiation (earlier it could have included interstellar matters) comes to the Earth unhindered. This period of feeble electromagnetic influence is believed to have duration of about 5000 years. It may, therefore, be logically assumed that in last 3500 million years more and more cosmic matters have accreted around the core of primordial Earth. Accretion of magnetized minerals in the core of the Earth under presently obtained thermal structure may not be possible because most minerals (in accordance with what is known as *Curie temperature*) would be melted by the depth of only 20 to 30 km from the Earth's surface. Elsasser, however, contended that movement of iron rich core consequent upon its affinity with the axis of rotation of the Earth would induce electric impulses continuously even if it has a wandering and weak magnetic field. It gives credibility to the existence of an outer liquid core surrounding the inner solid core.

**Temperature and Earth's internal structure:** Temperature distribution within the Earth is difficult to be assessed. However, based on the evidences collected from deep drills and laboratory experiments it may be suggested that thermal gradient and pressure conditions are closely associated within the Earth. It is found that the temperature in upper crustal layer increases with depth. Average rate of increase in temperature has been estimated to be 1° C for 32 meters of depth in upper crust of the Earth. It suggests that temperature may be increasing with a rate of 20° - 30° C for every kilometre of depth. But this rate of increase in temperature (thermal gradient) is not expected to continue because the temperature will be ranging between 800° C and 1200° C. at a depth of about 40 km. which is above melting point of most of the rocks found in lower crust and the upper mantle at this depth. If the same gradient is extended to about 100 km. depth it will be experiencing temperature magnitude of 2000° to 3000° C at which presently solid rocks in upper mantle would melt. If this rate of temperature increase continues temperature at the core of the Earth would be around 150,000° C. It would make the Earth unstable which it is not. It implies that the heat from the internal sections of the Earth must be flowing towards surface. Thus, extrapolation of temperature deeper down the Earth is not as simple as it is made out to be.

Investigations of rocks and laboratory experiments suggest that the rate of increase in temperature is closely related to the types of rocks and pressure they undergo during their phase change and which make different layers of the Earth, Geophysical studies reveal that the rate of temperature increase in upper layer of crust is higher than the lower layers. It is attributed to the fact that the radioactive minerals which abound in this layer up to a depth of about 100 km. are a major source of heat generation in this layer. As they are found to be lesser with increasing depth, rate of temperature increase declines rapidly. It is proven by the fact that the temperature of magma in volcanic vents with their base below 100 km. strata of partially melted rocks is recorded to be about 1200° C. On the basis of density differentiation of materials now it is estimated that the temperature in the core of the Earth may not be more than 5500° C.

**Pressure and Earth's internal structure:** Pressure within the Earth is also measured in terms of atmospheric units. One atmosphere unit is considered to have a pressure of about 6.68 kg. per 2.54 square centimetre (14.7 pounds per square inch). The pressure within the Earth increases due to overlying mass of materials. It is estimated that at a depth of about 1500 km. the pressure equals to almost 1 million atmospheres. It increases to almost 3.5 million atmospheres at the core. Pressure distribution within the Earth has a major influence on phase change of rocks and their properties including density e.g. Fe which has a density of 11 at the core has a density only of 7.8 at the surface of the Earth. These variations influencing the rock properties are also reflected in the behaviour of energies like seismic waves, magnetism, and heat transfer through them.

**Density and Earth's internal structure:** On the basis of such evidences geophysicists and geologists are now almost certain that the Earth comprises layers of matters of different composition and densities. The average density of the Earth, as determined by Henry Cavendish in the year 1798 and accepted by the geophysicists since is 5.52 g/cm3. Analyses of the surface rocks reveal that they generally have a density ranging from about 2 to 3.3 g/cm3. It underlines the fact that the Earth has an uneven distribution of density which in all likelihood goes on increasing with depth. It also reflects that the Earth is not composed of homogeneous matter. It may be speculated with reasoning that when the solidification of the Earth started some 4.6 mya (see module-4) its own field of gravity must have induced a process in which heavier and denser substances like nickel and iron accumulated towards the centre of the Earth and elements with larger proportion of silica and lesser density were arranged accordingly over the central core. As the Earth is spherical in shape materials with different chemical composition and specific gravity must have arranged themselves in concentric layers over each other.

Studies on temperature- pressure relationship suggest that rocks undergo definite phase change initially at depths of 100-250 km (lower crust- upper mantle boundary) followed by depths at 400 and 700 km.

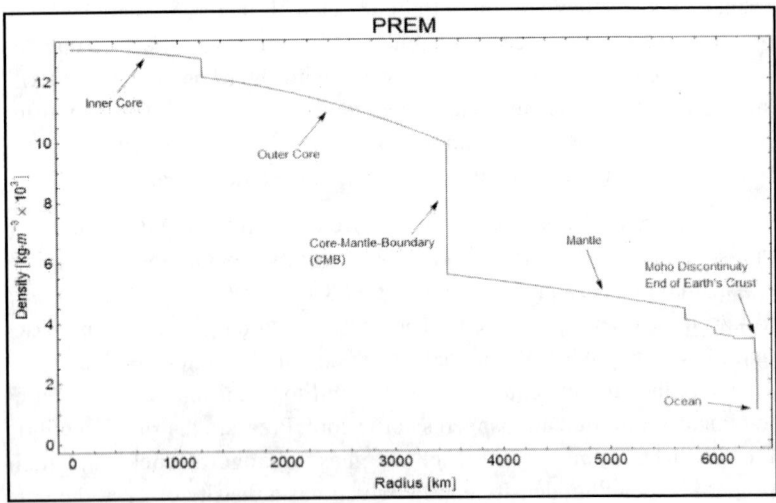

After Preliminary Reference Earth Model (PREM).

**Fig. 5.2: Earth's radial density distribution**

They also mark the changing density of rocks. Lower crustal density of 2.9 suddenly increases to 3.3 in upper mantle. Again a rapid increase in density of rocks takes place at the junction of partially molten upper and solid lower mantle at a depth of about 700 km. Thereafter, it is believed that density change in rocks is gradual till mantle-outer core boundary is reached at a depth of about 2900 km. Rocks again register a rapid change in their density from about 5.5 at the bottom of the mantle to 10 in upper outer core. Similarly, along outer- inner core boundary at a depth roughly of 5150 km. density changes rapidly from about 12.3 to 13.3. Maximum density of rocks at the centre of the inner core may be as high as 13.6. Abrupt density change from one layer above to the lower is known as discontinuity. A number of discontinuities have been identified through the analysis of seismic waves.

## 5.2: The Earth and Its Structure

Evidences collected from different sources have led to a general agreement that the Earth is composed of concentric layers of rocks having different densities and chemical attributes. There is also an agreement that density of these layers goes on increasing from upper layers to the core of the planet. Attempts have been made by scholars to identify different layers on the basis either of their compositional attributes or their strength and rigidity. It is considered useful to know these classifications as the names used in them continue to be contextually used even presently.

### Classification of Earth Layers by Suess

One of the earliest attempts to classify different layers within the Earth was made by **Suess**, an Austrian geologist in second half of the 19th century. He based his classification on the principal elements available in layers. He named them accordingly. He named the layers as (i) **Sial** corresponding with the upper layer (crust); (ii) **Sima** corresponding with the intermediate layer (mantle); and (iii) **Nife** corresponding with the core. The names are based on combination of chemical symbols of the elements dominant in each layer. Thus, Sial is symbolic acronym for Silicate and Aluminium; Sima for Silicate and Magnesium; and Nife for Nickel and ferrous (iron).

In view of Suess the upper layer (Sial) existed below sedimentary deposits of the continents. Rocks in this layer were dominated by acidic granite and granitic gneiss. Specific gravity of rocks in this layer varied between 2.75 at the top to 2.90 at the bottom. The intermediate layer (Sima) underlie the sial. It had a predominance of heavier basic plutonic rocks like basalt and gabbro. Specific gravity, according to him, varied from 2.90 near the upper crust to 4.75 near the internal layer. The core of the Earth or internal layer (Nife), he believed, was largely composed of heavy matters like nickel and iron with a specific gravity of 12.

## Classification of Earth Layers by Arthur Holmes

**Arthur Holmes**, a British geologist on the basis of his equation of heat loss from and heat generated within the crust by radioactive minerals attempted to arrive at thickness of different layers of the Earth. He concluded that the upper layer of Sial must extend to the upper part of what Suess named intermediate layer. In other words, there had to be a zone in which upper layer rocks with the dominance of Sial fused with the intermediate layer Sima. He called the upper layer the Crust having a transition zone in the intermediate layer which together could be named **Lithosphere**. With the Combination of data obtained from thermal composition of rocks, movement of seismic P- and L-waves, and initial depth of geosynclines he concluded that mean thickness of the crust could be between 15 and 30 km. where density would be ranging from 2.75 in the upper parts to 2.90 at the bottom. This crust rested over denser lower part of Intermediate layer of Suess. Holmes called it substratum and which also formed the part of the Lithosphere. He believed that the lower part of the Lithosphere extended up to a depth of about 70 km. and where it fused with the plastic property of the **asthenosphere**. His view on thickness of layers gives ample scope to explain landform features on the basis of active internal forces. In fact his Convection Current Theory takes into account these internal forces to explain landforms.

## Classification of Earth Layers by Van Der Gracht

Based on the information available in 1928, Van Der Gracht, a German geologist believed that the Earth consisted of four concentric layers. These layers had different composition. They also varied in their thickness. His four layers consisted of - (i) **'Outer Sial Crust'** composed mostly of felsic rocks (silicates of aluminium, potassium, sodium, and iron); (ii) **'Inner Silicate Mantle'** with a dominance of maffic rocks (silicates of magnesium, iron, and calcium) in which silicates of aluminium, potassium and sodium were scarce; (iii) **'Pallassite' zone**- a transition layer of ultramaffic rocks comprising mixed silicates of magnesium, iron and nickel; and (iv) **Nucleus or the Core** composed mostly of iron and nickel.

He, unlike Holmes, concluded that the average thickness of the outer Sial Crust under the continents was about 60 km. But under the oceans it was variable. Below Atlantic bed, he estimated that the thickness of this crust could be + 20 km as compared to continental blocks whereas it was totally absent in the Pacific Average density of this crust varied from 2.75 at the top to 2.90 at the bottom. An inner Silicate mantle was considered to underlie it and in which magnesium increasingly replaced

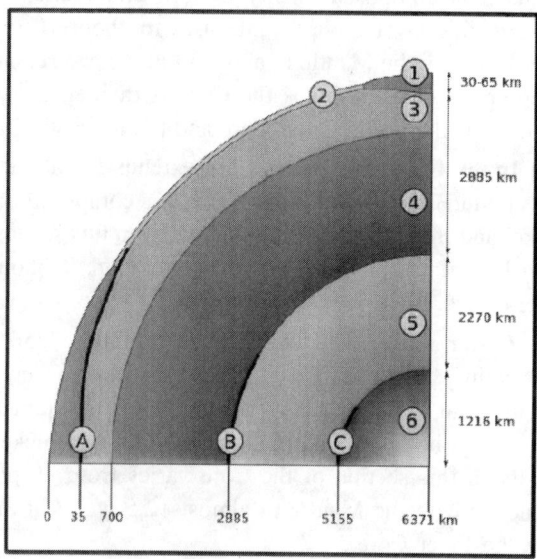

*Fig 5.3: Compositional Layers*

aluminium of the upper Sial crust. He believed that the average density of this layer at a depth of 60 km. was 3.1. But it increased to about 4.75 at a depth of about 1200 km. Below this layer was the 'Pallassite' zone which extended up to a depth of 2900 km. Density of rock at this depth was calculated to be 5. This layer was believed to surround the Core of the Earth extending from a depth of 2900 km. to 6371 km. Specific gravity of rocks in this layer was estimated to be 11.

On the basis of earlier studies it is obvious that the interior division of the Earth may be obtained keeping in view either (A) the compositional properties, or (B) the mechanical properties. Identification of compositional layers takes into account the chemical composition of different layers whereas mechanical classification takes into account rigidity and strength of the layers. It is mostly based on the behaviour of the seismic waves.

On the basis of chemical properties the Earth may be said to have broadly three major layers. They vary in thickness and compositional properties. Changes in elemental and thermal composition influence their density with depth. They are discussed below from the centre of the Earth to its upper most layer in that order:-

1. **The Core:** The Core occupies a little over half of the Earth's diameter. It occupies about 3471 km. of the inner parts of the Earth. It is composed mostly of iron and nickel. Density of rocks in this layer abruptly increases to about 10 g/cm³ at the boundary with the Mantle to about 14 at the centre. On the basis of compositional attributes the Core is divided in two distinct sections (i) the liquid Outer Core and (ii) the Solid Inner Core.

   **The Inner Core:** The Inner Core extends for about 1220 km. to its boundary with the Outer Core. It is composed of solidified Nickel and Iron (= Nife). Specific gravity in this section increases from 13.3 at its boundary with the Outer Core to about 14 g/cm³ at its centre.

   **The Outer Core:** The Outer Core is about 2250 km. thick surrounding the Inner Core. It extends from a depth of 2900 km. to about 5150 km. below sea level. It is in liquid state and is believed to be composed of liquefied iron and nickel. Specific gravity in this section of the Core varies from 10 g/cm³ at the boundary with the Mantle to almost 12.5 g/cm³ at the junction with the Inner Core.

2. **The Mantle:** The mental represents the intermediate layer of the Earth between the upper Crust and the internal Core. It extends up to a depth of about 2900 km. below sea level. It accounts for about 80% of the Earth's volume. It may be divided in two parts which may be designated as Lower Mantle and Upper Mantle respectively.

   **Lower Mantle:** It is composed of **ultramafic** silicate rocks (rocks composed largely of solid iron, magnesium, and silicon oxides) of high density. It is about 2230 km. thick. The Lower Mantle despite a very high temperature is in a solid state. It is attributed to intense pressure in this zone. Its density varies from 5.5 from the boundary of the outer core to about 4.3 to its junction with the Upper Mantle.

   **Upper Mantle:** the Upper Mantle is about 420 km. thick. It is believed to be composed mostly of maffic rocks with the dominance of viscous nickel. It is plastic in nature. Long duration stresses set in a lateral flow in this section of the layer. Specific gravity in this layer varies between $3g/cm^3$ in upper parts to $4.75$ $g/c^{m3}$ in lower parts. It is believed to have a specific gravity of about $4.3g/c^{m3}$ at the boundary with the Lower Mantle to $3.3$ to that with the lower Crust.

3. **The crust:** The crust represents the upper most layer of the Earth. It is composed mostly of silicate rocks that are rich in elements like aluminium, potassium, sodium and metallic ferrous. They have a low density ranging between 2 and $3.3g/cm^3$. Even iron which is estimated to have a density of $11g/cm3$ at the core of the Earth has a density only of $7.8g/cm^3$ in this layer. There is a general agreement that this layer has a thickness of 5 km under the oceans. But it may be 35 to 70 km. thick under the continents. They, however, may be identified to have two distinct types as described below:-

**Continental crust** is variable in thickness and composition. Upper part of this layer though not continuous has a thin layer of sedimentary rocks which may be extending to a depth of 35 km. (60 km.and 70 km. according to Gracht and Holmes respectively) under the folded mountains. They are mostly composed of felsic to mafic elements having an average density of $2.65g/cm^3$. It is underlained by crystalline granite and gneiss. Below them basaltic rocks are found. These rocks of igneous origin are sometimes exposed on the surface of the continental shields.

**Oceanic crust** is uniform in thickness and composition. It is about 8 km thick. Rocks in this layer are mostly mafic in nature with increasing proportion of magnesium. They are composed mostly of igneous rocks (basalt and gabbro) overlain by thin layer of sediment.The differences in thickness and density between continental and oceanic crusts allow the relative stability of land features floating over a denser mantle (**see chapter on Isostasy**).

As is obvious, these compositional layers and sub-layers have very sharp and abrupt boundaries. It also requires to be remembered that the presence of liquid outer core is responsible for the presence of magnetic field around it and which provides it a protective cover against most of the cosmic incidences.

## 5.3. B: Mechanical Layers

Identification of layers has also been made on the basis of rigidity and strength of different segments of the Earth. They have mostly been recognized on the evidences arrived at by the analysis of meteorites and behaviour of passing seismic waves. These layers do not necessarily correspond with the compositional layers. The Earth may be divided in seven major mechanical layers with many minor ones in them. Many Mechanical Layers particularly the lower ones retain the names given to the Compositional Layers. As important evidence with regard to mechanical layers is obtained from the behaviour of earthquake waves, which originate only in the upper parts of the Earth, layers here have been named from top to the inner most. They, thus, may be identified as **(i) Crust. (ii) Lithosphere, (iii) Asthenosphere, (iv) Upper Mantle (v) Lower Mantle or Mesosphere, (vi) Outer Core,** and **(vii) Inner Core**

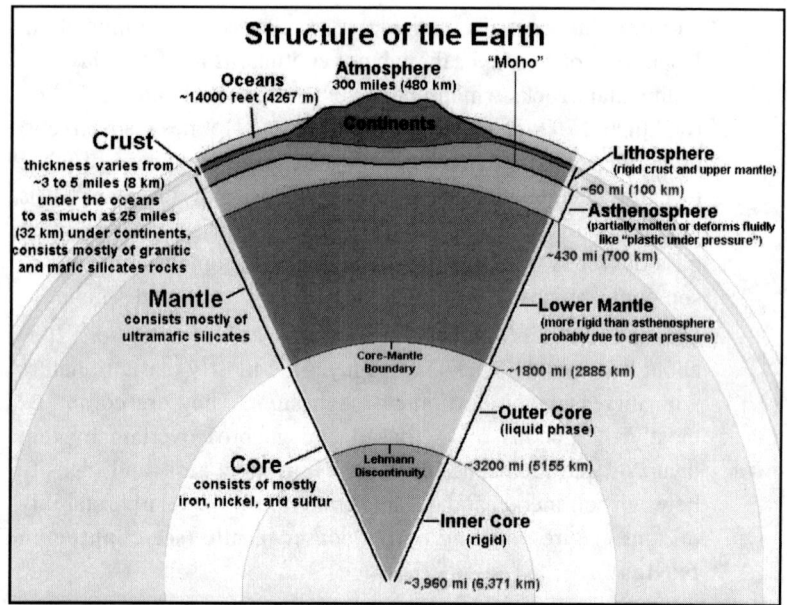

***Fig 5.4: Earth's Mechanical layers in relation to the Compositional layers***

(i) **The Crust:** The Crust is the upper most part of the Lithosphere. It is most rigid of the layers. They represent only 1% of the Earth's structure. They may be divided in two categories **Continental** and **Oceanic**.

(i) Continental crust is generally less dense composed of granitic rocks. On the other hand the oceanic crust is basaltic and denser. Continental crust has an average thickness of 40 km where as oceanic crust is only 8 km. thick. Only under mountains the continental crust may be as deep as 35 km. Seismic velocity which is determined by the density of rocks is the least in this layer. In addition, only in this layer L- waves or surface waves can be identified.

(ii) **The Lithosphere:** The lithosphere forms the outer most layer of the Earth. In fact, the Crust is also a part of it. It is the most rigid section of our planet. It is different from the Crust in the sense that it includes a thin section of Mantle also. It is believed to extend up to a depth of 60 to 150 km. It is thicker under the continental blocks and thinner below deep ocean basins. It is separated from the underlying Asthenosphere by a structural anomaly known as **Mohorovicic discontinuity**. Popularly this discontinuity is referred to as **Moho**. Seismic waves are found to be accelerating below this discontinuity as denser matters are encountered.

(iii) **The Asthenosphere:** Below Lithosphere lies Asthenosphere. It generally is found below a depth of 60 km. under the oceanic lithosphere and 150 km. under the continental. It may be considered the upper part of the Upper Mantle. It differs from the lower parts of the Upper Mantle in the sense that it contains scattered zones of radioactive decay which generates high temperature. As a result, the matters become less rigid and viscous in this layer. In context of seismic waves it is also known as **Low Velocity Zone.** In fact, the asthenosphere is a transition zone separating the strong, solid rock of the uppermost mantle and the Crust above from the remainder of the strong, viscous parts of upper mantle below. Most of the crustal deformation processes, earthquakes, and volcanoes owe their origin in this layer. Having a molten characteristic matters assume convective motion which is believed to accommodate displacements in overlying crustal lithosphere.

(iv) **Upper Mantle:** Under the Asthenosphere separated by another discontinuity is the main body of the Mantle. It extends from about 250 km. to a depth of 700 km. It is in a semi liquid state. Therefore, seismic waves relatively lose their velocity in this layer.

(v) **Mesosphere** or the **lower Mantle:** This part of the mantle is solid despite a very high temperature. Within this layer also two minor structural anomalies have been identified at a depth of about 1700 and 2450 km. These anomalies are reflected in refraction of the seismic waves.

(vi) **The outer core:** This layer extends for about 2250 km. below the Mesosphere. Temperature in this part is much higher than what it is in Mesosphere. As a result, iron and other elements are liquefied in this layer. This leads to slowing down of P- waves and complete absence of S- waves beyond this layer. The anomalous structural condition at its boundary with the Mesosphere is known as Gutenberg discontinuity.

(vii) **The inner core** or the **Barrysphere:** The Inner Core has a radius of about 1220 km. from the centre of the Earth. It is solid. As a result, seismic P-waves move with relatively greater velocity and accelerate again in this layer. The separation zone between the Outer and Inner Core has been named as Transition (Lehman) discontinuity. Internal structure of the Earth as should be obvious from the above discussion has a definite bearing on its surficial features. It is, thus, directly linked with spatial arrangements that are essential for the growth of human civilization.

## Review Questions

1.   Compare and contrast the major layers of Earth.
2.   Describe the nature of the Earth's inner core and its characteristics.
3.   Describe the two types of crust found on Earth. What is the basic composition of each type of crust?
4.   List the most common elements in Earth's crust.
5.   Identify the major zones of Earth from the centre to the surface. How do these zones differ from each other? What is the special significance of the asthenosphere?
6.   Define the differences between continental crust, oceanic crust, lithosphere, and asthenosphere
   1.   Describe the Earth's inner structure, from the centre outward. What types of crust are present? How are they different? What is the *moho*?
   2.   Compare and contrast the oceanic and continental lithosphere.
   3.   Briefly describe the density stratification of the Earth.

4. What is the difference between a P- wave and S- wave?
5. What is the low velocity zone?
6. How do we know that density of rocks in the Earth generally increases downward? Why do topographically high mountains have a negative Bouger anomaly?
7. What is heat flow? How is the heat flow in the Earth related to thermal gradient?
8. How does seismology help in understanding the internal structure of the Earth and what are its main findings?
9. Give a reasoned account of velocity variations of seismic waves in the interior of the Earth.
10. Write short notes on the following:-

    (a) Sial and Sima, (b) Low velocity zone, (c) Thermal state of the Earth's interior, (d) Moho (e) Magnetic field of the Earth, (f) Discontinuities within the Earth, (g) Mesosphere, (h) Asthenosphere

## References

1. Burchfiel, B.Clark, Robert J. Foster, Edward A. Keller, Wilton N. Melhorn, Douglas G. Brookins, Leigh W. Mintz, Harold V. Thurman (1982): Physical Geology; Charles E. Merrill Publishing Co.; London.

2. Christopherson, Robert W. (2013): Elemental Geosystem (7th ed.), Pearson Education, Boston.

3. Dey, A.K. (1968): Geology of India; National Book Trust, India; New Delhi.

4. Sharma, R.S. (2009): *Cratons and Fold Belts of India*, Lecture Notes in Earth Sciences 127, DOI 10.1007/978-3-642-01459-8_2, C_Springer-Verlag Berlin Heidelberg 2009.

5. Strahler, A.N. (2011): Introducing Physical Geography (5th edition); John Wiley & Sons; New Jersey.

# Chapter - 6

# Isostasy

## Objectives

Main objective of this chapter is to make students understand how the land features maintain gravitational equilibrium i,e, state of balance on differentiated layers of a mobile Earth. It aims at investigating the relationship that exists between surface features and the substratum. And though the Principle of Isostasy has been accepted universally its operation is shrouded in uncertainty. This chapter, therefore, aims at examining the following:

The antecedence leading to the development of this Principle,

Hypotheses that have been advanced to suggest operational mechanism for isostatic balance.

Geological processes and Isostatic adjustments.

## 6.1: Concept of Isostasy

The concept of Isostasy derived from two Greek words Iso + Staios literally means 'in equipoise' or in a state of balance. It is a geographical phenomenon that refers to the force of gravity acting upon crustal materials of different mass and volume in order to keep them afloat. It is believed that below certain depth in the interior of the Earth the total weight per unit area has to be the same all around the earth. It requires to be understood that Isostasy is not a process or a force. It is a principle which attempts to uncover the natural adjustment or balance in different segments of the Earth. It underlines the fact that higher crustal features e.g. continents or mountains which are composed of less dense matters protrudes proportionately below in order to maintain a stable gravitational equilibrium. The depth below which gravitational equilibrium is maintained, has been called the "**depth of compensation**". The concept of isostasy, however, is concerned only with vertical segments of the Earth and their movements. It is this characteristic that made **Steers** (1937: 70) to observe that 'any hypothesis on the causes producing the surface structure of the earth and demanding serious consideration must take into account the doctrine of isostasy'.

The concept, in general, takes into account the natural balance of mass that exists in the three upper layers of the Earth- (i) The Crust, (ii) The Lithosphere, and (iii) The Athenosphere (discussed in chapter-5). The term 'isostasy' was used for the first time by **C.E. Dutton** in 1889. He used the term to suggest 'gravitational equilibrium' existing between crustal features and which evidently are having different mass and dimensions.

Concern about the Principle owes its origin to a geodetic survey in the Andean region undertaken by **Pierre Bouger** in 1735. During the survey he realized that the Andes were not attracting the plum bob which a uniformly dense and voluminous feature like it should have affected. His finding did not evoke much interest then. But the problem was revisited near the Himalayas during the course of geodetic survey of India. **George Everest**, the Surveyor General of India, while determining the latitude of Kalianpur and Kaliana about 603 km. apart on the same longitude (70°E) observed that latitude established through astronomical method was lesser by 5.236 seconds of an arc than that of obtained through triangulation in

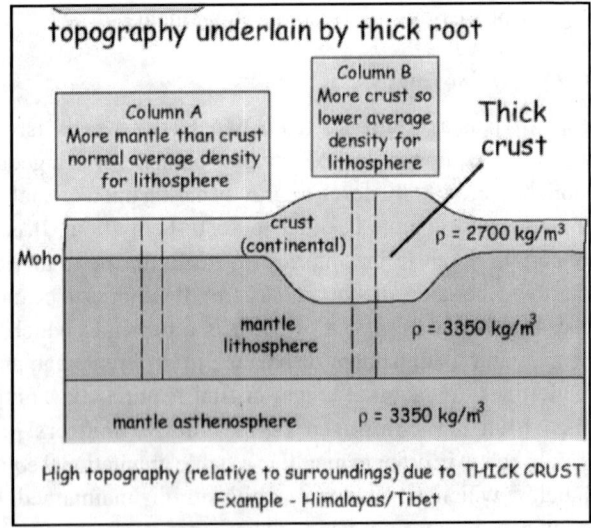

*Fig. 6:1 Representation of Airy's view*

which such anomaly would not occur. In other words, readings obtained by the two methods differed by 108 meters in length between the two places. The difference was attributed to greater deflection of the plum bob (used in astronomical survey) at Kaliana which was located only about 90 km, from the Himalaya. It was later discovered by Pratt, Arch deacon of Calcutta and a mathematician that despite its bulk the deflection was not proportionate near the Himalayas. On the other hand, the survey in Indo-Gangetic Plain to the south of the Himalayas suggested a southward declination of plum bob and where no surface feature could be compared with the Himalayas. The anomaly suggested that the density of rocks underlying central India was higher than those of the Himalaya. The mystery evoked much interest among the geologists and geophysicists since middle of the 19th century.

A number of hypotheses have been proposed to explain the mechanism by which higher surface remain stable on a mobile Earth of varied compositional layers and why some areas with lower features appear to have greater specific gravity? Deficiency or excessiveness of proportional attraction by land features referred to as **'Bouger anomaly'**, after the person who encountered the problem first.

## 6.2: Isostasy Hypotheses

### Root Hypothesis of George Airy

**Airy**, a British astronomer, was one of the first to address the problem of mass deficiency amongst the earth features. Following the findings in India he hypothesized in 1855 that the density of the Crust ($2.75g/cm3$) was same from top to bottom. They, however, had variable thickness. He, in order to explain the equilibrium amongst the surface features opined that the Crust was a rigid mass which floated over viscous and denser Mantle below. He in tune with the **'Law of Buoyancy'** also suggested that the higher parts of the crust proportionately sank down more into the mantle as compared to the lower parts of the crust. It implied that the continental blocks had deeper projection in the Mantle than the oceanic crusts which were thinner. It also meant that greater penetration of mass of higher surface features proportionately displaced the denser matters below in order to maintain **hydrostatic equilibrium.** It, however, did not mean that all features would displace the underlying denser materials individually. It was suggested that the Crust had sufficient strength to bear the weight of individual and isolated features above the surface. It was later calculated by **Hayford** that an area more than one square degree

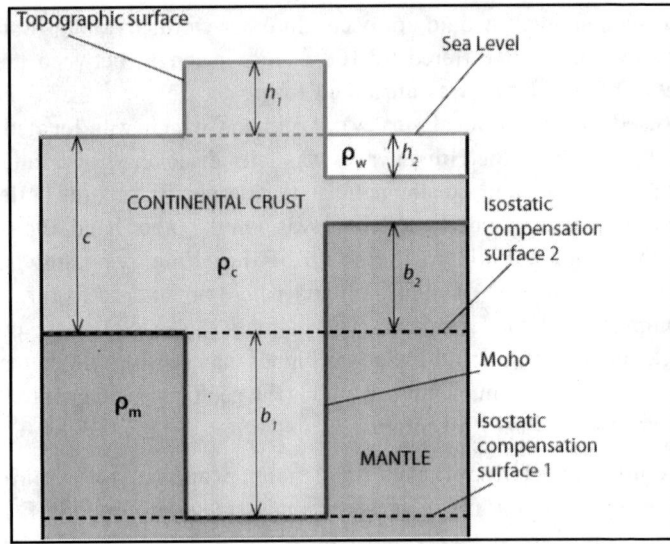

*Fig. 6.2a: Illustration of Airy and Pratt's view*

of the Crust could only impact displacement of denser materials below. Airy's idea of compensation of higher surface features by proportionate displacement of denser materials below (*variable thickness and uniform density*) came to be known popularly as 'Root Hypothesis'.

## Level of Compensation of John Henry Pratt

**Pratt**, a contemporary of **Airy** and living in India, provided a different mechanism to explain the gravitational equilibrium amongst the earth features. He also believed that the continental crust and associated features stood higher than the average because they were composed of less dense materials. On the other hand, the surface composed of denser materials represented depressions and hollows in the form of oceans and oceanic crust. His calculations of the difference in deflection of plum bob at Kalianpur and Kaliana showed it to be more than the observed deflection. It meant to him that the Himalaya was not attracting the plum bob as much as its mass was expected to. This led him to suggest that there existed an inverse relationship between altitude of surface features and their density. He, on the basis of his finding, also concluded that low density of higher crustal features was the result of expansion of matters due to heating at the bottom. It was in consonance with the calculation of **Joly** that there was complete liquefaction of matters at a depth of about 100-110 km. Pratt, therefore, believed that there would be loss of density at that depth without disturbing the volume of the surface features. Thus, he concluded that there was a He calledlevel below which solid crust melted but with the same density as features above the sea level. In simple terms, higher features over ground were compensated by less dense matter below them. And despite their varying elevation above the sea level they would maintain single level at the base in substratum having uniform density. He unlike Airy, therefore, believed in *'uniform depth with varying density'*. The figure.6.2 shows the basic difference in the view points of Airy and Pratt.

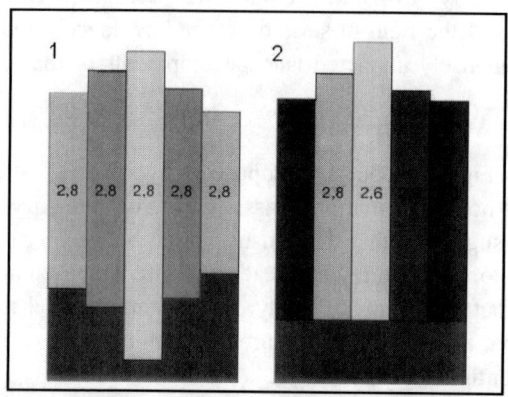

*Fig. 6.2b: Viewpoints of Airy and Pratt*

## Zone of Compensation: Views of J.F.Hayford[1] and J.W. Bowie

Hayford and Bowie, the two American geodesists supported the view of Pratt. They were first to bring the principle of isostasy from the domain of assumptions. They were able to prove that the surface features and gravity anomaly were intrinsically related and thus, validated the idea of isostasy. They are also accredited to suggest the depth of isostatic compensation. Hayford was also the first to give an estimated depth of isostatic adjustments. In order to place the depth of the 'plane of compensation' they assumed that the density was laterally uniform below certain depth in the interior of the Earth. Variation in the density of the crust and its association with elevation, they believed, was found only above this depth. Keeping in view, however, the underlying extent of crust in tune with the general law of buoyancy (which suggests that the 8 unit of the Crust sinks in underlying substratum for each 1 unit above it) and the thermal structure in zone of crustal penetration Hayford later modified his view. He, instead, visualized a zone of compensation having a thickness of about 15 km. below the continents and in which the density variations were neutralized to facilitate isostatic adjustments. It was suggested that where the zone was lighter a down ward projection of the higher surface features takes place. It, in a way, was the affirmation of Airy's view albeit in a limited zone. But Joly did not agree with this concept of 'zone of compensation'. He thought that the temperature at the base of a 100 km thick Crust, as assumed by Hayford, would be so high that all matters will completely liquefy. There will not be any density variations hence any isostatic adjustments cannot take place in this zone.

It requires to be mentioned that when the concept of isostasy was developing many information were not available to its pioneers. Advancement in the field of seismology and evidences obtained through them has immensely impacted later developments of the concept.

### (D) View of W.A. Heiskanen

Heiskanen, a Finnish geodesist, proposed a hypothesis that is considered essentially to be a combination of assumptions of both Airy and Pratt. He unlike them suggested that the density varied between crustal blocks as well as from top to bottom within each of them. He based his assumption on the fact that the average density of rocks at sea level is found to be more than that at high crustal features. It implies that the differentiation in density continues deeper down.

Heiskanen's assumption appears to be logical in the sense that increasing temperature with depth with additional pressure on lower portions of the Crustal column would bring about density change. As different blocks are believed to have different densities they extend

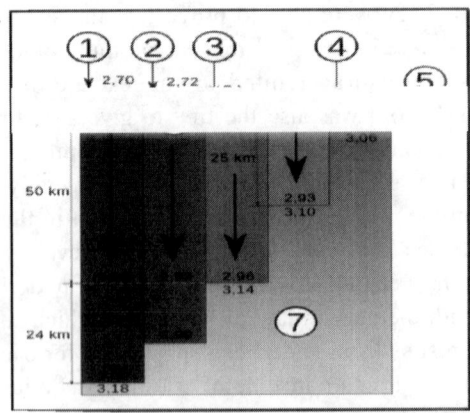

*Fig. 6.3: Lower density more depth*

proportionately to different depths. Now it has been established that the upper crustal rocks have density of 2.67 g/cm³. It increases to 2.90 g/cm³ in its lower parts forming the upper Lithosphere and which extends to a depth of about 70 km. At this depth density of rocks is suggested to be 3 g/cm³. Seismic evidences also reveal that there exist at least 2-3 minor discontinuities in density structure till the major Moho discontinuity at a depth of about 70 km. separates Lithosphere from Asthenosphere. It is found that the density of rocks changes from 3.3g/ cm3 in the lower Lithosphere to 4.3 g/cm³ at a depth of 250 km. where Asthenosphere mingles with the Upper Mantle. Asthenosphere being viscous and plastic in nature provides room for isostatic adjustments of higher surface features as lateral movement of matters is possible in this layer.

**Heiskanen** hypothesis, therefore, meets the objections raised against his predecessors in respect of proportionate projection of crustal rocks in the substratum. Thus, there appears to be a broader greater acceptance of **Airy**'s view on isostasy.

**Holmes** (1926) also subscribed to this point of view. In fact, **Jaffreys** (1929) commenting on isostasy also felt that "the small gravitative effect of mountains is explained on Airy's lines as due to squeezing out of a weak but dense substratum by the weight of loads added on top. It is shown that compensation distributed uniformly down to a definite depth is consistent with Airy's mechanics, but that with the actual structure of the Crust compensation is probably concentrated at the base of the granitic and intermediate layers" (quoted in **Steers**; 1937:75).

### 6.3: Geological Processes and Isostaic Adjustment

There are quite a few geological processes that may be attributed to continually going on isostatic adjustments below the Lithosphere. They may briefly be outlined as follows:

The heat generated in the interior of the Earth and its dissipation checked by the overlying solid Crust goes on increasing. It introduces a molten condition facilitating generation of convection currents in Upper Mantle. Colliding convection currents are likely to drag the lower parts of solid crust deeper down and melt them. It creates geosynclines (huge structural depressions) on continental surface.

Sedimentation in these geosynclines through sub aerial processes and their increasing weights further deepens them. As a result, floors of these huge depressions gradually fuse with the underlying magma. Magma rises from the weakened basements to the surface as **volcanic eruption**. Where they are not able to reach above the surface magma intrusion leads to the formation of **batholiths**. These activities release the heat and convection currents die out.

## Consequently the crust rises and erosion processes set in again.

Convection currents in the interior of the Earth, though very slowly, induce movement of the crustal blocks. They are considered responsible for the movement of the continental plates leading to their disintegration in one phase and aggregation in other. There are evidences of at least 4-5 such incidences in the geological history of the Earth since late protorozoic eon to the present.

Erosion processes which unload sediments of higher altitude on the surface make them lighter. They rise in the denser substratum with denser materials below them. On the other hand, where deposition of sediments takes place load increases and that portion sinks in substratum. This mobility removes the denser matter under them proportionately. Thus, equilibrium is established between emerging and submerging masses.

It is the unloading of ice sheets from Pliestocene glacial epochs that has uplifted many parts of continental blocks as in Scandinavia, Greenland, Alaska, Baltic coast and Antarctica. It is evident in distinct features like sea-cliffs and wave-cut platforms which are presently seen hundreds of meters above present sea level. The process of uplift of land mass caused by removal of load on the top is termed **'rebounding'**. The process is more 'probable in mountain ranges bounding plateaus where erosion and removal of surface materials are more intense due to unique drainage patterns. A good example is provided by the Tibetan Plateau bounded by the Himalayas and Kunlun Mountains (Ollier and Pain; 2000).

It is found that chemical transformation at the roots of high rising features may make them denser. It may lead to their detachment in the Mantle. They may be moved by the underlying convection currents in the Asthenosphere. Detachment of the roots provide Asthenosphere room to rise. The Cenozoic uplift of Sierra Nevada region is attributed to this cause (Boyd *et al; 2004)*. This may also be the cause behind the vertical instability of the Himalayas which may have its root foundering under the Eurasian plate.

It may be concluded from the above discussions that isostasy is a simple idea. And though its impact on tectonic movements is not yet fully established it provides significant explanatory power in relation to the vertical structure of crustal features and the interior composition of the Earth.

## Review Questions

1. What is isostasy?
2. What are the unusual gravity measurements?
3. What are the models of isostasy?
4. Is the Himalayas under isostatic equilibrium?\
5. What are the isostatic effects of sedimentation and erosion?
6. What is gravity?
7. How is gravity measured?
8. Differentiate between Level of Compensation and Zone of Compensation.
9. Write short notes on:-
   • Bouger anomaly
   • Rebounding
   • Batholiths
   • Isostatic adjustment
   • Hydrostatic equilibrium
   • Law of buoyancy
   • Declination of Plum bob

## References

1. **Boyd, O. S.; Jones, C. H.** and **Sheehan, A. F.** (2004): "Foundering Lithosphere Imaged.
2. Beneath the Southern Sierra Nevada, California, USA," Science 305, 660.
3. Burchfiel, B. Clark, Robert J. Foster, Edward A. Keller, Wilton N. Melhorn, Douglas G. **Brookins, Leigh W. Mintz, Harold V. Thurman** (1982): Physical Geology; Charles E. Merrill Publishing Co.; London.
4. **Dayal, P**. (1990): A Text Book of Geomorphology; Shukla Book Depot; Patna.
5. **Gilluly, J**. (1970): "Crustal Deformation in the Western United States," in *The Magatectonics of Continents and Oceans*, ed. by H. Johnson and B. L. Smith (Rutgers,), p. 47.
6. **Holmes, A,** (1965); Principles of Physical Geology; Nelson; London Ollier, C. and Pain, C. (2000): *The Origin of Mountains;* Routledge.
7. **Steers, J.A.** (1937): The Unstable Earth: Methuen & Co. Ltd.; London.
8. **Strahler, A.N.** (2011): Introducing Physical Geography (5th edition); John Wiley & Sons; New Jersey.
9. **Wadia, D.N.** (1975): Geology of India (4th ed.); Tata-McGraw Hill Pub. Co. New Delhi.
10. **Watts, A. B.** (2001): Isostasy and Flexure of the Lithosphere; Cam

## Note

1. **Hayford** was also the one who proved that the Earth was not exactly a sphere rather it was an oblate sphere.

# Chapter - 7

# Crust Materials: Minerals and Rocks

## Objectives

Earth Crust, the home of innumerable organisms including man, is basically composed of inorganic substances. Interaction of it with three sub-systems- the atmosphere, the hydrosphere, and the biosphere; makes the Earth a living planet with ever changing dispensations. An understanding of interplay between the processes and Earth materials and its resultants is expected to enable one to explain man's milieu both past and present. This, therefore, necessitates a rational understanding of materials of which the solid exterior of our Earth is composed and the processes which form them and bring about changes in them. In the light of the fact that Man is the ultimate beneficiary of inherent rock properties an attempt has been made in this chapter to briefly give an account of basic earth materials under the following subheads:

Minerals and Rocks; nature and classification

Processes of their formation and Rock cycle;

Impact on man.

## 7.1: Minerals and Rocks

**Minerals:** Minerals are substances that have particular chemical composition and well defined arrangement of elements in them. It is noteworthy that the earth materials considering their weight are formed mostly of eight (8) of 112 known elements. Most dominant of them are Oxygen, Silicon, Aluminium, Iron, Calcium, Sodium Potassium, and Magnesium. They roughly constitute 99.2% of the weight of the earth materials. Rest of the elements together have a negligible presence of only about 0.8% in them (See fig.7.1).

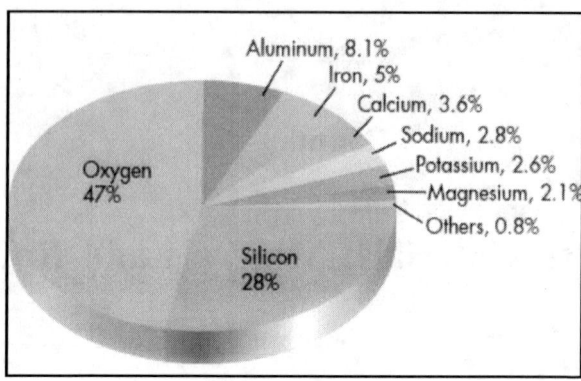

After Arbogast (2014: 310)

**Fig. 7.1: Proportion of chemeical elements in Earth crust.**

Most common mineral forming elements are formed due to differential synthesis of silicon and oxygen with other elements. Distinctiveness of chemical composition and physical characteristic of minerals depend on the arrangement of elements within them. Arrangement of elements in minerals may go through rearrangement and pressure of elemental temperature molecules under different conditions.

Thus, the minerals may undergo phase change. It may alter their chemical properties as well as physical form. A new mineral comes into existence distinct from the parent mineral. There are about 4200 minerals. But about 30 of them are most common and which may be cemented together either individually or with other minerals to form rocks.

Minerals of the earth crust are distinguished from each other on the basis of alignment and proportion of the elements that make them. Minerals may also be distinguished on the basis of level of crystal formation. Crystals in mineral are formed in the process of cooling and their solidification from a molten state. It refers to an orderly arrangement of atoms that make elements. The process of crystallization in minerals with the exception of mercury (liquid matter cannot have definite internal order) is controlled by numerous processes and underlying natural factors.' *Minerals*', in the light of above mentioned characteristics, may, therefore, be defined as '*naturally occurring crystalline, inorganic substances having a definite range in chemical **composition and physical properties**'* (**Burchfiel** et al. 1982:38). These properties are manifested in their hardness; crystal forms; fracture and cleavage; colour and streak; luster (reflectivity); taste and solubility as well as in their specific gravity. Difference in these attributes make minerals distinct from each other.

## 7.2: Identification of Minerals

**Hardness**: One of the methods of identifying minerals is by establishing their hardness. It is made on the basis of *'hardness scale'*. Hardness scale has been developed taking into account the hardness of ten minerals. The scale is known as *Mohs scale*. Ranking of these minerals has been made from the softest to hardest. They accordingly have been ranked from **1** to **10** in ascending order of hardness. Talc, the softest of the minerals is assigned a hardness of **1** followed by Gypsum, Calcite, Fluorite, Apatite, Orthoclase, Topaz, Corundum and diamond in that order.

**Crystal Form**: Form of crystal is another indicator by which minerals may be differentiated. Depending on the space and time available during the process of mineralization generally seven system of crystal are recognized. They are *Isometric* (e.g., Diamond), **Hexagonal** (e.g., Vanadinite), **Tetragonal** (e.g., Scheelite), **Orthorhombic** (e.g. Topaz), **Monoclinic** (e.g., Brazilianite), **Triclinic** (e.g., Labradorite) and **Trigonal** (e.g., Rhodochrosite). If solidification of minerals takes longer time crystals in them accordingly become larger and may be seen with naked eyes. If solidification is rapid crystals are proportionately small and minute. Some of them may be observed only with the help of microscope or X-ray. In other words, size of crystal in different minerals depends on time of cooling and solidification of elements in molten matters. Normally crystal size in minerals is found to be proportional to temperature and pressure distribution in segments of the lithosphere. It may be remembered that heat and pressure successively decrease in lithosphere from bottom to its upper layer (see Chapter -5).

**Fracture and Cleavage**: One of the significant parameters for recognition of minerals is the manner in which they break. Irregular disintegration of minerals is called fracture. Fractures may have many patterns. They may be **smooth** (as in *mica*), **fibrous or splintery** (as in *asbestos*) or **choncoidal** (circuitous break as in *quartz* or glassy *osbedian*).

On the other hand, many minerals break along a plane of weakness and expose the lower plane. These planes created by weak bonding in crystalline minerals represent cleavage as in mica, galena, and calcite. These cleavages may have six directions in common minerals (dolomite, calcite etc.). **Colour:** Reflectivity of minerals also distinguishes one mineral from the other. Colour of minerals suggests its capacity to absorb light which by implication refers to their chemical bonding e.g. colour of **graphite** may range from black to steel gray; that of **hematite** from reddish brown to black or gray whereas **sulphur** exhibits a yellow colour).

However, colour of minerals, particularly of non-metallic ones, depends on the level of its purity e.g. **diamond** in its pure form is White but it may have pinkish, bluish, or greenish look if it is dented with some impurities. Similarly, **quartz**, **garnet**, **topaz** etc. may be found in many colour sheds.

More accurate identification of minerals may be made with the help of **streak**. Streak represents the colour of powdered minerals and which many a time differs from the colour of minerals in lump. When a mineral despite its colour variation is rubbed on hard plain surface it produces constantly a definite colour line. These lines are referred to as streak. **Graphite** and **magnetite**, for example, make black streaks; **hematite** streak varies from light to dark red; **Chalcopyrite** has greenish black streak whereas **sulphur**, **diamond**, **quartz**, **garnet** etc. do not produce any streak colour.

**Lustre:** Lustre refers to brightness of minerals. It depends on their reflectivity. Different types of minerals have different lustre. Minerals may broadly have two types of lustre **metallic** and **non-metallic**. Metallic lustre may either be **bright** like *pyrite or galena* or **dull** like *chalcopyrite*. Non-metallic lustre may be grouped under self explaining terms like **adamantine** (brilliant lustre like *diamond*), **vitreous** (glassy like *quartz*), **resinous** (like *opal*), **greasy** or **waxy** (like *serpentine*), **pearly** (like *talc*), and **silky** (generally of fibrous minerals like *some varieties of asbestos*) and **earthy** (as of *hematite*).

**Taste and Solubility:** Minerals also vary in taste and solubility. These characteristics of minerals depend on the nature of elemental precipitates that combine to form them. A soluble mineral may have tinge of sweetness where as insoluble one may have glassy feel on tongue. Solubility of minerals may also be tested with the help of diluted **hydrochloric acid**. Powdered minerals react with it which is reflected in formation of bubbles on its surface. Reactions may vary from weak to very strong depending on the chemical composition of minerals.

**Specific gravity:** Specific gravity of minerals refers to the relative weight of mineral making substances to that of equal volume of water. All minerals vary in their specific gravity depending on the mineral constituting elements. It is generally found that that minerals formed in the interior of the Earth have higher specific gravity than found in the crust (*see chapter- 5*).

# Rocks

Lithosphere of the Earth is basically composed of Rocks. They are *those earthly matters which are formed due to aggregation of one or more minerals*. As mentioned earlier there are about 30 minerals which are foundational to rock formations. These minerals singularly or in combination are the building blocks of rocks. All rocks, however, have a predominance of quartz that belongs to silicon dioxide family of minerals. Despite the fact that rock forming minerals are limited various rocks forming process affecting mineral combinations produce diverse types of rocks. Oldest of the rocks of the Earth are believed to have been formed some 4.4 billions of years ago when the Earth was still in a process of cooling *(see Chapter-4)*. Rock formations since have been continuing. Even presently when molten magmatic materials are ejected from the volcanoes they solidify as rocks. They are continuously being formed at mid-oceanic ridges. In addition, all earlier formed rocks are exposed to several sub aerial agents and endogenic processes. They disintegrate and are deposited in layers to form other generation of rocks. Pre existing rocks as well as derived rocks may also undergo transformation under certain heat and pressure. These processes induce changes in their cohesiveness and attributes. Sometimes changes brought in by different forces may entirely alter the physical properties of derived rocks. Genetically, however, all rocks may be grouped under three major categories viz. (i) **Igneous**, (ii) **Sedimentary**, and (iii) **Metamorphic**. Of these rocks igneous rocks are the first to have been formed in initial

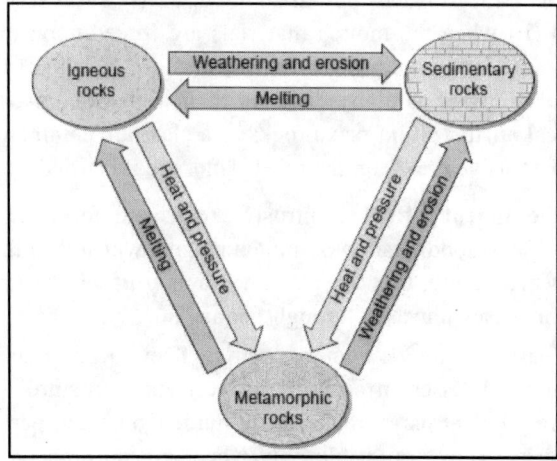

After Strahler (20011: 376)

***Fig. 7.2: Rock Cycle***

stages of Earth's solidification. All other type of rocks, directly or indirectly have been derived from them. Igneous rocks, therefore, are known as **Primary Rocks.** Following figure describes the processes under which different groups of rocks are formed and recycled.

## Igneous Rocks

The word Igneous has been derived from Latin word 'Ignis' which means fire. Igneous rocks are formed due to solidification of molten magmatic materials and volcanic lava. Almost 65% of the Earth Crust comprises these rocks or their fragments. In fact, most of the secondary rocks would have a very high proportion of materials incorporated in them. These rocks are considered to have been formed earliest in the geological history of the Earth. All other type of rocks, directly or indirectly, owes their formation to them. They, therefore, are named as **'Primary Rocks'.** These rocks consist of silicate minerals dominated by bonding of silicon and oxygen. They also contain metallic elements. Minerals in these rocks are very tightly interlocked. Presence of quartz which is resistant to chemical reaction makes them very compact, impervious, strong and resistant.

## Classification of Igneous Rocks

Igneous rocks may however, be further classified on the basis of their closely linked mode of occurrence and texture. These rocks form broadly in two ways:- (i) When the molten sub lithospheres f matters solidify below the surface they are known as **(i) Intrusive Igneous Rocks;** and (ii) when the molten materials are forced upon the surface and solidify (as during volcanic activities) they are called (ii) **Extrusive Igneous Rocks.** They naturally differ in their mineral composition and other physical attributes. Intrusive rocks as a class are commonly known as **diorite.** Extrusive rocks, on the other hand, are identified as **andesite.**

**Intrusive Igneous Rocks:** Intrusive rocks are formed below the surface due to solidification of molten magma which is in the form of **plutons.** Therefore, they are also known as **plutonic rocks.** Magma or plutons rises upwards through certain openings.

These passages are known as **vents.** They occur due to some inherent weaknesses in sections of overlying structure. Naturally, therefore, the intrusive rocks are bounded on their periphery by older rocks. They are visible only when erosion processes remove the overlying layers in which intrusion has taken place. When intrusion cuts across the overlying structure it is known as **discordant** intrusion. Many intrusions, however, may not be able to make passage through overlying strata of rocks. Instead, magmatic matters make room between weaker sections of systematically laid stratums. In such cases intrusion is said to be **concordant**.

**Discordant intrusion:** Largest of the discordant igneous intrusion is represented by **batholiths** which may be hundreds of kilometre wide and thousands of meters deep. They are found to be underlying most of the mountain systems of the world. Smaller intrusions in overlying rocks are known as **stock.** As solidification process is very slow at great depths due to heat and pressure, cooling of matter is extremely slow. Cooling of matters and crystallization is also dependent on the nature of minerals that combine to form these rocks. In upper parts of intrusion it is generally **felsic** minerals (abundance of *light coloured and less dense silicate minerals with low melting temperature*) which compose these rocks. A good example of such rocks is **granite** which develops below a depth of about 1.6 km. Mineral constituent of rock further downwards is dark and denser in their composition. They are known as **mafic** minerals. Mafic minerals have a high proportion of relatively denser magnesium silicate. **Gabbro** is composed of mafic minerals. Still deeper down mafic minerals may have two heavier elements in them. They are termed **ultramafic** minerals. **Peridotite** is one example of ultramafic rocks. Generally, however, intrusive rocks are less dense than the extrusive rocks. In any case, it takes a very long time for these minerals to crystallize. As a result, coarse and visible crystals develop in these rocks.

Another type of discordant intrusion of igneous rocks is represented by **dikes.** Dikes are formed when upcoming magma forces to open vertical fractures. It occupies those straight and almost parallel cracks in the crustal rocks. They form tabular bodies which may vary from a few centimetres to some thousand meters in their width. Generally, however, they are found to have width ranging from few meters to tens of metres. Their depth is greater than their width. They may have a depth of many kilometres. Due to their resistant nature they may be exposed as ridges when overlying matters are removed by sub aerial agents of erosion. As dikes are much smaller than batholiths and have much more surface area relative to their volume, cooling of matters is relatively quicker. Thus, crystals formed in them are accordingly smaller and finer.

**Concordant intrusions**, on the other hand, refer to the intrusion between sedimentary stratums. They are identified as **sills** and **laccoliths.** Sills are very thin and are unable to bring about visible deformation on the surface. They have fine grained texture. But laccoliths are substantially voluminous and generally lead to up-warping of overlying surface. Depending on their volume crystal grains in them may be coarse or fine.

**Extrusive Igneous Rocks:** Rocks formed due to solidification of molten magma on or near the Earth's surface are called extrusive igneous rocks. Volcanoes and other volcanic activities eject molten magma known as lava on the surface. Lava in contact with atmosphere and hydrosphere cools and solidifies very quickly. As a result, very little or no crystallization takes place in these rocks. These rocks are mafic and contain dissolved gas and little amount of silica. They are viscous and glassy in nature. Basalt is a good example of extrusive rock. Rocks, thus, obtained are also called **volcanic rocks.** Figure 7.2 provides a generalized view of mode of formation of different igneous rocks

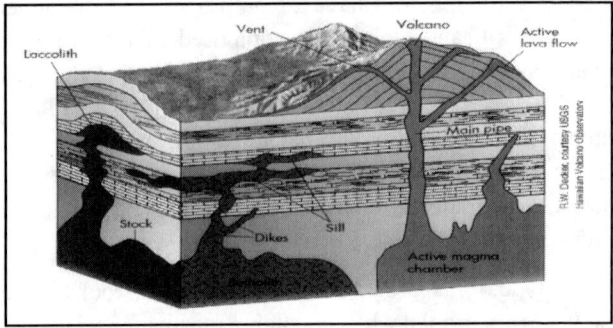

*Fig. 7.3: Formation of Igneous Rocks.*

| Granite | Diorite | Gabro | Rhyolite | Andesite | Basalt |
| Felsic | Mafic | Felsic | Mafic | Felsic | Mafic |

INTRUSIVE    EXTRUSIVE

From Arbogast, (2014: 311-312)
*Fig. 7.4: Classes of Igneous rocks and their characteristics*

## Sedimentary Rocks

Sedimentary rocks represent the second major group of rocks found on Earth. They constitute about eight percent (8%) of the crustal rocks and almost 70% of the surface rocks. They are formed at or near the Earth's surface. They are obtained from all type of rocks (igneous, sedimentary and metamorphic) though ultimate source is igneous rocks. These rocks are invariably impacted by **denudation** (*erosion, transportation, and deposition*) operated by sub aerial agents. Erosion of rocks liberates innumerable particles from the parent rocks which are transported by different agents as **sediments**. Broken particles and unconsolidated minerals i.e **detritus** thus obtained is deposited in strata in suitable environments as **sedimentary deposits**. Rounded fragments of rocks and loose minerals are known as **clasts**. Weight of overlying deposits in all types of rocks forces some common minerals like calcium and silica to recrystellaized in sediments. Recrystellaized minerals act as glue. Loose minerals (sediments), thus, are cemented and become compact to form sedimentary rocks. The process of reconsolidation of matters is known as **'lithification'**. Few of the sedimentary rocks though may have lithification of organic matters most of them have inorganic origin as the parental source of sediments ultimately is igneous rocks. Lithification transforms sediments into solid and recognizable layers. The process of solidification and strata formation vary in their nature as cementing materials  effectiveness would be different in different environments (fluvial, arid, glacial, karst and marine). As a result, different sedimentary rocks exhibit different physical properties. **Sedimentary rocks, thus, are those rocks which form from deposition and** lithification mostly of fragmented and dissolved inorganic substances of other rocks and organic residue of terrestrial and marine organisms.

An understanding of the nature of different stratum of these rocks and their attributes gives an insight in geological history of the Earth. Fossils are preserved only in sedimentary rocks. Critical evaluation of these rocks enables one to understand temporal changes in respect of land water distribution and palaeo climate. Study of sedimentary rocks acquire more significance because most of the economically important minerals are found in these rocks,

Classification of Sedimentary Rocks
Sedimentary rocks may be categorized in two groups-
* **Clastic sedimentary Rocks**, and
* **Nonclastic sedimentary Rocks.**

**Clastic sedimentary Rocks:** Derived from Latin clastus the word clast means broken or fragmented. These rocks are formed from broken parts of an earlier rock. Rock fragments generally acquire a round shape during denudation processes. Rocks of this group are classified on the basis of size of the clast. They range from small to very large. They are generally identified on the basis of texture. Following table (7.1) gives name of some common clastic rocks with their normal physical attributes.

**Table 7.1: Some Common Types of Clastic Sedimentary Rocks and their characteristics**

| Rocks | Composition |
|---|---|
| Conglomerate | Coarse grained sandstone containing rocks and pebbles of various sizes |
| Sandstone | Sand grains cemented by minerals |
| Siltstone | Silt particles cemented by minerals |
| Clay stone | Clay particles cemented by minerals |
| Shale | Fine particles containing fossils; split along layers; Most abundant |
| Limestone | Calcium carbonate obtained from both inorganic as well as animals and microorganisms |

From Arbogast (2014:314)

**Nonclastic sedimentary Rocks:** These kind of sedimentary rocks based on their mode of formation may be categorized in two groups-

Rocks of chemical precipitates, and

Rocks from organic deposits.

**Rocks of chemical precipitate:** Chemical precipitates from minerals are common in sea waters and salt lakes. It is associated with precipitation and recrystallization of calcium carbonate in saline waters. There are many factors which may activate these processes. Generally, however, it may be the rise in sea water temperature, high evaporation, upwelling of water and bacterial decomposition amongst other factors which set in the process of precipitation and recrystallization as well as their lithification.

On the other hand innumerable calcium rich organisms are found in marine environment. After they are dead they are deposited on the sea floors. In course of time they are cemented by the mineral precipitates to form rocks. Most of the **lime stones** rich in calcium carbonate have been formed like this. If there is predominance of magnesium in mineral precipitates of calcium carbonate i.e. calcium magnesium carbonate they form **dolomite.** These rocks are generally rich in fossils which are indicative of the environment in which organisms evolved. Besides, sedimentary rocks also form on surface due to evaporation of water leaving behind salt residue and simultaneously resulting in crystallization of sodium. These are called evaporites. Following table (7.2) gives some example of sedimentary rocks formed on account of mineral precipitates.

**Table 7.2: Rocks formed of chemical precipitates under seawater and salty lakes**

| Rocks | Composition |
|---|---|
| Dolomite | Magnesium carbonates formed by chemical replacement in limestone |
| Evaporites | Minerals (gypsum. rock salt. calcite) residue after evaporation of water |
| Chert or Flint | Very fine grains of silica (quartz) in layers or nodules within limestone |

*Source:* Author's own From Arbogast, (2014:314)

**Rocks of organic deposits:** Rocks of organic origin are formed due to thick deposits of decayed plants and animals. These rocks are generally formed in an environment which is acidic and water logged. Such a condition checks oxidation with proportionate increase in carbon content. It helps in initial solidification of these deposits in the form of peat. Buried under weight of progressive sedimentation over millions of years and due to increasing pressure and heat water and oxygen contents are squeezed from these carbonaceous deposits. Under compaction they lithify and are converted into coal.

Organic sedimentary rocks generally occur in layers. These layers are mostly separated by intermediate layers of sandstone or limestone or shale. Microscopic plants and organisms under these conditions liquefy to form crude oil. Decomposition of these organic substances is a major source of natural gas containing large proportions of methane and ethane with small quantities of propane and butane. Table-7.3 lists some of the important rocks of organic origin with their significant characteristics.

**Table 7.3: Important Organic Rocks and mode of their occurrence**

| Rocks | Properties |
|---|---|
| Coal | Carbonaceous materials, Lithification under heat and pressure |
| Crude Oil | Liquid hydrocarbon trapped in sedimentary deposits |
| Natural Gas | Gaseous hydrocarbon trapped above that of liquid form in sedimentary deposits |

From Arbogast, (2014:314)

## Metamorphic Rocks

A very large part of the earth crust is composed of metamorphic rocks. They account for about 27% of the Earth crust. In fact, lower parts of the solid lithosphere are believed to be formed of this category of rocks. Metamorphic literally means 'changed form'. Metamorphic rocks, therefore, refer to changed forms or transformation of pre-existing rocks commonly called **'protolith'**. These rocks may be igneous or sedimentary or metamorphic rocks themselves. Besides undergoing relocation of constituents under denudation processes, rocks of the Earth are regularly subjected to transformation due to increase in pressure of overlying load of rocks and consequent rise in temperature underneath. Similar changes may also be seen if rocks come in contact with fluids having dissolved chemicals. The process (es) sets in a change in arrangement of physical properties and chemical bonding in rocks. It brings about changes in their texture, colour and density. Rocks, thus, acquire traits which are significantly different from the original ones. The process affecting the change is known as **Metamorphism**. Metamorphosis of parent metamorphic rocks is referred to as as **'Progressive Metamorphism'**.

Metamorphism, however, is a process which affects only solid rocks before they are liquified. Under this process mineral components are rearranged and recrystallized. Type of metamorphism depends on the nature of energy that is responsible to bring about change(s) in the parent rocks. As mineral compositions of rocks vary the critical temperature and pressure causing change in them are different. Generally the process of transformation in rocks is set in when temperature exceeds a range of 150- 2000C as well as when they come under pressure above 1500 bars. These variables may beget change either singularly or collectively. On the basis of processes inducing changes in rocks three types of metamorphism has been identified. They are named as (i) *Thermal Metamorphism*; (ii) *Regional metamorphism*; and (iii) *Dynamic*

**Thermal Metamorphism:** It is also known as contact metamorphism. It refers to the process when rocks on margin come in contact with intrusive igneous bodies like those of batholiths, laccoliths, dykes etc. High temperature at the time of intrusion of magma in these bodies also brings about thermal change in rocks which are in close proximity to them. Mineral component adjust accordingly in these rocks to change their original properties. Metamorphosed rocks obtained in this fashion are called **hornfels**.

**Regional metamorphism:** Metamorphism under this process takes place over a very large area and millions of year. It is generally induced by crustal movements. Crustal movements may be caused by increase in weight of collected sediments in broad depressions and shallow seas present on the surface of the Earth. It is closely associated with mountain building phases. Sediments in depressed zones under its own weight create sufficient heat and pressure on its own underlying sections to transform their original properties. Nearness to a very hot and viscous upper mantle also adds to the increase in heat. Squeeze under pressure and increasing temperature with depth leads to **metamorphosis** of rock forming materials. As is obvious, this kind of transformations in rocks occurs at great depths. The rocks that are metamorphosed under this process show distinct orientation of new minerals in them consequent upon their critical temperature of recrystallization. These rocks are, therefore characterized by foliation in them and are known as **foliated** rocks. **Gneiss** and **Schist** are the most common foliated metamorphic rocks.

**Dynamic Metamorphism:** This type of metamorphism is not very common. It is set in by large scale movement of crust along some fault systems of the crust in which rock masses are thrust over others. Such movements lead to enormous compression of rocks when they come in contact with each other. This leads to disintegration and reshaping of rock forming materials due to friction. The process generally does not affect crystals of the parent rocks. A very common example of this kind of metamorphic rocks is **cataclasites** and **mylonites.**

Bases of all the mountains comprise metamorphic rocks. Occasionally they are exposed when crustal cover is removed over them by sub areal agents of erosion or they are uplifted during mountain building episodes.

## Classification of Metamorphic Rocks

Metamorphic rocks are generally classified on the basis of the process of their transformation as discussed above. They may, however, be classified broadly in three categories which may transgress the processes involved in their formation. They are-

- Nonfoliated Metamorphic Rocks.
- Foliated Metamorphic Rocks, and
- Dynamic Metamorphic Rocks

**Nonfoliated Metamorphic Rocks:** This type of rocks is characterized by their contact with intrusive rocks i.e they develop as thermal metamorphic rocks in narrow belts. In this condition grains of minerals in them which are almost invisible may or may not be recrystallized. If their crystals (irrespective of size) are re-formed they generally lose their original characteristics entirely. On the other hand, they may only undergo slight adjustment of mineral constituents with respect to their size. In that case most of the characteristics of original rocks are preserved in the transformed rocks. In the following table metamorphosis of these two different types of nonfoliated rocks is shown.

**Table-7.4: Metamorphic rocks from nonfoliated rocks**

| Nonfoliated Parent Rocks | Obtained Metamorphic Rocks |
|---|---|
| Any Sedimentary Rock | Hornfels (re crystallized) |
| Sandstone (clastic rock) | Quartzite (no change in crystals except the size) |
| Limestone (non-clastic) | Marble (no change in crystals except the size) |

From Arbogast, (2014)

**Foliated Metamorphic Rocks:** Most of the rocks formed due to regional morphogenesis show structural foliation. Forces involved in their formation are almost similar to those making mountain structures. These rocks exhibit lamination and striped texture. Stripes in these rocks give a clue to the direction of force that was exerted on them. Increasing metamorphic processes lead to grade variations in the same parent rocks. Following table provides an idea about the grade of metamorphism in different metamorphic parent rocks.

**Table-7.5: Changing Grades of Metamorphic Rocks**

| Parent Metamorphic Rock with foliated texture | Low-Grade metamorphism | Medium–Grade metamorphism | High -Grade metamorphism |
|---|---|---|---|
| Shale (Metamorphosed from Clay) | Slate | Schist | Gneiss |
| Basalt Extrusive (Igneous) | Greenschist | Amphibolite | _____ |
| Granite Intrusive (Igneous) | Slate (if felsic) | _____ | Gneiss |
| Limestone & Dolomite | | | |

From Arbogast, (2014)

**Dynamic Metamorphic Rocks:** This type of rocks refers to the rocks obtained of metamorphosis of broken or grinded rocks generally found in narrow belts of fracture on the crust. As they are found very near the surface they are neither exposed to high temperature nor much pressure. As a result, no recrystallization takes place in them. Instead, they may have all grades ranging from coarse to fine grained of rocks. If cohesion in this type of rocks is weak they are known either as **fault gouge** or **fault breccias**. If they have greater compactness of similar rocks they are called **mylonites**. These type of rocks, however, are generally recognizable only in areas of their occurrence

## Rocks and Mankind

Rocks since the early periods of the evolution of human civilization have been considered a valuable resource. Survival of mankind and indicators of civilization since early days has been closely related with how rocks and materials obtained from them. It is believed that man developed his first armament against hostile animals some 2.5 m.y.a. There are evidences that sturdy **rock caves** provided them with first shelter against unbearable sun, unbearable winds and torrential rains. It has been the level of efficiency (generally associated with technology) to produce and use **weapons, shelter, utensils, instruments, jewelleries** which separated groups of people from one another. It is obvious in the fact that those early civilizations are referred to as **Palaeolithic (Stone Age)**. Later Stone Age represented by the term **Neolithic** suggests better and improved methods of production and handling of equipments derived from stones. Over the time with the discovery of **flint rock** fire making could be controlled. It, apart from meeting the household requirement and protection against wild animals, also facilitated smelting of rocks to get stronger yet more flexible minerals to combat adversities of the environment man lived in. It all started with the discovery of techniques to get metal from rocks popularly referred to as **Bronze Age**. It culminated in development of smelting iron from rocks. It marked the advent of **Iron Age** ushering additional power and probably more security as durable artefacts could be moulded out of it. Stability in living condition encouraged them to look for more usable from rocks. Metallic ores of gold, silver etc. were found and used. Ultimately, they became the symbols of wealth and prosperity. Man learnt to use charcoal as a source of controlled energy. The Post Industrial Revolution developments expanded the use of inanimate sources of energy like coal, crude oil, natural gas and radioactive minerals. All of them are either rocks themselves or essentially derivatives of rocks.

Presently no aspect of human life is untouched by rocks and their products. With the expansion of information technology industry demand for conducive minerals like silicon is increasing leaps and bound besides others.

Besides meeting the simpler demands of 'terra cota' industries bulk mining of sand and clay provides globally the basis for one of the most economically lucrative sector-the Real Estate and is least wasteful. Wastage in mining of Precious minerals like copper and gold is found to be very high generally beyond the resilience of the environment. For example, in order to recover 1 ounce of gold about 12 tonnes of ore is wasted. In case of copper 30 tonnes of rock materials have to be wasted for procurement of one tonne.

According to the World Bank estimates about 23 billion tonnes of net mineral are mined every year at global level. In order to obtain them about 50 billion tonnes of overburdens are removed. The total materials removed are believed to be equivalent to the deepening of Earth surface to the tune of 1.5 metres year after year. Mining activities in formal sector, though accounts for only about 30 million (about1%) of the total workforce of the world, estimated 6 million workers are suggested to be engaged actively in small mining mostly under unorganised sectors and which supports almost 300 million of their dependents. Contribution of mining in economy at global level though is very small it provides bulk of economic sustenance to many of the technologically backward developing nations particularly in Africa, parts of South America and Asia. As a matter of fact, most of the manufacturing and transformation industries are directly dependent on this wealth of the Earth. 10 million out of the total 30 million miners are engaged only in coal mining which continues still to be the major source of energy for all types of industries. Coal contributes to more than 80% of the global energy requirements.

India by volume has emerged as the 4th largest producer of minerals in the world. In terms of value the country ranks 8th in the community of world nations, The country, based on 2013 study, mines and processes about 87 minerals in which are extracted 4 sources of fuel, 3 atomic minerals and 80 non fuels. Of non fuels 10 metallic minerals, 47 non-metallic minerals and 23 micro minerals are mined. Yet contribution of mining as an independent economy is only 2.2 to 2.5% to its GDP. If entire industrial sector is included its share would be somewhere between 10 and 11%. It gives employment to about 700,000 persons in working age group.

Apart from the economic benefits of minerals its unrestrained exploitation and uses have brought to the fore the perils glaring at the humanity as whole. Even if one overlooks the industrial disasters since second half the last century, environmental degradation caused by mining and uses of products is becoming unmanageable threatening the survival of the planet. In whatever manner one likes to interpret there is no denying that rocks are intrinsically linked with the survival of the mankind present and future.

## Review Question

1.  What is a mineral? Name the most common minerals on Earth. What is a rock?
2.  What is the difference between intrusive and extrusive igneous rock? Give few examples of each kind of rock. Explain the properties that differentiate them?
3.  What are the various kinds of sedimentary rock? Explain lithification in context of the formation sedimentary rocks.
4.  What is metamorphism, and how are metamorphic rocks produced? Name some original parent rocks and their metamorphic equivalents.
5.  Describe interchanges that occur within the rock cycle. Which rock is involved in the beginning of cycle? Explain the process transforming igneous rocks to sedimentary rocks?
6.  In what conditions sedimentary rocks become metamorphic rocks?
7.  Describe igneous processes. What is the difference between intrusive and extrusive types of igneous rocks?
8.  Characterize both coarse- and fine-grained textures; how is this related to the cooling history of the rock?
9.  How do the processes of erosion, deposition, and lithification fit within the rock cycle?
10. How is our economy dependent on organic sedimentary rocks and why are we vulnerable as far as petroleum resources are concerned?
11. What is meant by an order of relief? Give an example from each order.
12. Explain the difference between relief and topography.
13. What are the major differences between folding and faulting? What causes these types of crustal activity?
14. Explain compressional processes and folding, and *describe* four principal types of faults and their characteristic landforms.
15. How do intrusive processes and volcanism differ from each other?
16. Diagram a simple folded landscape in cross section, and identify the features created by the folded strata.
17. Define the following terms: igneous intrusion, pluton, batholith, and dike.
18. What is a **fault scarp**?
19. Describe and explain the formation of landforms that result from normal faulting (such as **grabens**, **horsts**, and **tilted fault-block mountains**).

20. There are basically two different forms of tectonic activity. These are _____ (a) compressional and extensional; (b) stressful and decompressional; (c) decompression and extensional; (d) compressional and stressful

21. Compression leads to the folding of the crust, which results in the formation of _____ (a) anticlines and synclines; (b) synclines and troughs (c) upfolds and troughs; (d) troughs and anticlines

22. _____ faults result in crustal shortening produced by compression of the crust. (a) Normal; (b) Transcurrent (c) Strike-slip; (d) Reverse

## Reference

1. Arbogast, A.F. (2014): Discovering Physical Geography; John Wiley & Sons; New Jersey.

2. Burchfiel, B.Clark, Robert J. Foster, Edward A. Keller, Wilton N. Melhorn, Douglas G. Brookins, Leigh W. Mintz, Harold V. Thurman (1982): Physical Geology; Charles E. Merrill Publishing Co.; London.

3. Christopherson, Robert W. (2012): **Geosystems : an introduction to physical geography**, 8th ed. Prentice Hall, New Jersey, USA.

4. Christian D. (20II): Maps of time: an introduction to big history; University of California Press, Berkeley.

5. Dayal, P. (1990): A Text Book of Geomorphology; Shukla Book Depot; Patna.

6. Gabler, Robert E., Petersen, James F., Trapasso, L. Michael, Sack Dorothy (2009): Physical Geography, Ninth Edition; Brooks/Cole, Cengage Learning; Belmont, USA.

7. McKnight, Tom L. and Hess D. (2011): McKnight's Physical Geography: *A Landscape Appreciation* (11th edi); Pearson Education, Inc.; Delhi.

8. Ollier, C. and Pain, C. (2000): *The Origin of Mountains;* Routledge.

9. Strahler, A, (2011): Introducing Physical Geography; John Wiley & Sons; New Jersey.

10. Wadia, D.N. (1975): Geology of India (4th ed.); Tata-McGraw Hill Pub. Co. New Delhi.

# Chapter - 8

# Tectonic Processes and Land forms

In foregoing chapter an attempt was made to explain the formation of materials that compose our Earth. This part is designed to give an insight in the process of evolution of landscapes and the forces that construct or deform them. It will also attempt to associate different forces with different land forms and ever changing landscape. This chapter, therefore, aims at the study of the following:

- Earth's primary reliefs : Permanency of Land and Ocean Basins
- Earth movements and Rock Deformation

## Earth's primary reliefs

Relief features are the topographic expressions on the Earth's upper crustal layers. Many forces either singularly or in tandem have been operating to form and deform the low density rocks since the solidification of the crust which is believed to have started some 4.6 billion years ago. As a result, reliefs of different dimensions and magnitude have been developing throughout the geologic history of the Earth. One finds vast oceanic basins and continental land masses at global level. Each of them is seen to be interspersed with the presence of vast features like mountain chains, plains and lowlands which many times have pan continental/ transnational expanses. These features may also comprise variations within them in the form of hills, hillocks and depressions or features like them at much smaller scales. Based on the scale of their occurrence, therefore, relief features of the Earth are generally grouped under three orders as follows:

Reliefs of First Order;
Reliefs of Second Order; and
Reliefs of Third Order

**Reliefs of First order:** Continental land masses and the oceans of the world possessing distinct topographic characteristics represent relief of the first order. The land masses all together cover about 29% of the surface area of the Earth. They, besides the continents, include all the big and small islands. Rest of the 71% area is covered with big oceans. Average thickness of continental masses is considered to be about 35-36 km whereas some of its sections may be protruding in the asthenosphere up to a depth of 70 km. These land masses also comprise the oldest solidified igneous and metamorphic rocks known as **continental shields** and around which they have been growing for billions of years (*see chapters 5 and 6*). Additionally, the average thickness of land though is measured from above the sea level a substantial portion of it is found to be extending along the coastlines of the continents below sea level. Basically the **Continental Shelves** and the **Continental Slopes** composed of sediments obtained from wear and tear of the adjacent land masses may only be considered their extension below the sea level. Figure 8.1 shows the distribution of continents and oceans proportional to their elevation and depth from the sea level.

**Ocean basins**, on the other hand, are natural depressions on the upper crust of the Earth below average sea level. Their bottom structure differs in composition from each other. For example, deep oceanic plains of the Indian and the Atlantic oceans are found to have some thin layers of about 2 to 3 km. dominated by **Sial** compounds. But this layer is almost entirely absent in the deepest of oceans- the Pacific. Like the continental masses most of them also possess distinct relief features like system of hills, plains and depressions. Some notable features of the continents and the ocean basins as they are found today are out lined below:

Oldest continental rocks are believed to have been formed some 4200 m.y.a. whereas the oceanic crust is not more than 200 m.y.a.

More than 39% of the northern hemisphere is covered by land. It accounts for more than 75% of the total land area of the world. On the other hand, southern hemisphere has only a little over 18.5% of land and 81% of big water masses. It is in this light that the northern hemisphere is called *Land Hemisphere* and the southern hemisphere is called the *water hemisphere*.

The Continents and the Oceans broadly are triangular in shape- Continents having their bases generally in circum Arctic belt occupied by Eurasia and North America. It is best reflected by the two

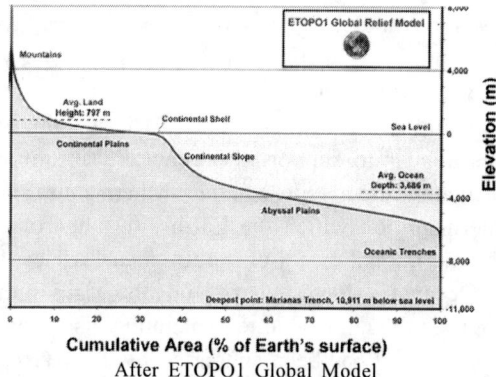

**Cumulative Area (% of Earth's surface)**
After ETOPO1 Global Model

***Fig. 8.1: Distribution of Continents and Oceans***

Americas. They together may be regarded as making an elongated 'isoscles triangle' with the Arctic Ocean as its base and Cape Horn its apex. The two American continents even independently are more or less triangular in shape. North America may be said to have its base in circum Arctic land mass with apex in isthumus of Panama. Similarly, South America may be visualized to have its base in Northern Brazilian Plateau and apex at cape of Horn. Though not as clear as in western hemisphere Eurasia may be said to form the bases of two triangles with their apexes in Africa at Cape of Good Hope and Tasmania respectively.

Northern Hemisphere though is dominated by land masses its Polar area is mostly occupied by *Arctic Ocean-* a water mass surrounded by circum polar land masses. On the contrary, the Polar region in Southern hemisphere is occupied by land mass of *Antarctic* surrounded by large circum polar water bodies.

Land and water masses are generally situated antipodal. A close look at a globe reveals that almost all the land masses big or small are diametrically opposite to an ocean. The only exceptions are Patagonia and New Zealand which are found opposite to parts of North China and Iberian Peninsula respectively. An attempt to explain this disposition of the Earth was made by **Lowthian Green** in 1875. He proposed *tetrahedral hypothesis*. Green's hypothesis was based on simple facts of elementary geometry which suggests (i) *spherical shape* allows to contain largest volume of surface area, and (ii) subjected to equal pressure during its multi factor cooling process the spherical shape would assume a *tetrahedral form* (a tetrahedron would have the least volume of the surface area). As a tetrahedron comprises faces and coigns, a tetrahedral Earth would have the continents (composed of lighter materials) gathered along the coigns from the very beginning with oceans occupying the faces. This naturally would make them antipodal. At first sight, arrangement of continents appear to support the view as Laurentian; Baltic and Siberian shields seem to represent the northern coign and the Antarctic the southern one on which the Earthly tetrahedron is believed to stand. The meridional extent of the Americas, Africa, and south-east Asia with Oceania is suggested to occupy the edges of the tetrahedron. The hypothesis though is able to explain the distribution of land and oceanic bodies and has been supported by **J.W. Gregory** in recent times; it has been rejected on the basis that the gravitational equilibrium of the Earth would not allow a spherical form to assume a tetrahedral shape or vice versa. His explanation once acclaimed by Arthur Holmes as ingenious is presently of only historical interest.

More recently studies in plate tectonics have been able to provide some plausible explanation to the development and distribution of these first order relief features of the Earth (*see chapter-9 on Mountain Building and Plate Tectonics*). All continental masses have grown around one or more initially solidified crust known as shields or table lands. Indian sub- continent is found to have grown around *peninsular table land* and which is bounded by the Indian Ocean on three sides and separated from Alpine-Himalayan system by the Indo-Gangetic-Brahmaputra plain in the North. **Africa** and **Australia** may be regarded as two stable land masses with minor abrasion in North West represented by the Atlas Mountains in Africa and eastern highlands of Australia supposedly the part of **epi-continental sea** of Palaeozoic era. The ancient shield of **Brazil-Guiana** forms the core of South America with a boundary with the Andes in west, hill provinces of Argentina in south and the mountains of Brazil in the east. South America may also be said to have another old land mass represented by the *Patagonian* region to the south of the Brazilian shield and bounded in the west again by the Andes. **Antarctica** is another of the shield which is surrounded on all sides by massive **Southern Ocean**. All these shields are generally considered to be parts of Gondwanaland.

Similarly, parts of northern super continent of Angara land or Laurasia Europe is found to have grown around **Baltic Shield** bounded in south by the Balkans, Carpathians and Caucasus mountains, in north-east by Ural Mountains, and in north and north-west by Caledonian and Hercynian Mountains. **Siberian Shield** is believed to lie between the

Urals Mountains in the west, hills of Taimyr Peninsula in north and Verkhoyansk Mountains in the east. The **Chinese Table** is broadly surrounded by the Island arcs in the east, Yunnan faults in south and west and marine deposits found north of Mongolia (*Siberian and Chinese Table lands together form northern part of Asia*). **Canadian Shield** is bounded in north by Arctic Archipelago, in south by mountains of early, middle and lower Pre-Cambrian periods and in the west by Rockies of tertiary origin.

Most of the big islands are found to be nearer the continental bodies mostly within a distance of 200 km. They are generally located on the continental shelves so have close Structural affinity with the continents. They are identified as **sub-continental islands**. Greenland, British Isles, Sri Lanka, Madagascar are some of the examples of sub-continental islands. There are many other islands which located far away from the land masses and rise above the ocean surfaces. They are called **Oceanic Islands**. Structurally they are different from the sub continental islands as sedimentary rocks in them are usually absent. Besides these two types of islands some islands are found in groups mostly along the marginal seas off the coasts of continents as **Island arcs**. They may have mixed structure. Some may be continental in origin and some may owe their origin to volcanic eruptions as in island arcs along the border of East Asia in western Pacific.

## Permanency of Continents Ocean Basins

Development of first order relief features and characteristics of their distribution have been intriguing since long. This essentially raised the question about the origin of continents and the ocean basins. Many hypotheses have been put forward to explain them. Many viewed that higher parts representing continents and oceanic depressions formed very early in the process of contraction of the Earth. Some like **Sollas** attributed them to differential initial atmospheric pressure. Primordial molten materials, according to him, were depressed in areas of high atmospheric pressure to be occupied by water masses later. The areas where atmospheric pressure was low remained relatively higher and formed continents. **Chamberlin and Moulton** attributed formation of continents and oceans to differential aggregation of planetesimals (discussed in Chapter 2:2:2). **Lowthian Green** (discussed earlier) attributed the distribution of the First Order Relief to changing earth forms. These hypotheses though may have some merits keeping particularly in view the time when they were proposed lack wide acceptance. Their significance lies only in the fact that they paved the way for later scholars to provide better explanation with respect to configuration and distribution of these major features of the Earth.

With growing information about the ocean morphology since the publication of Alfred Wegener's works between 1912 and 1929 on displacements of continents there has been a plethora of evidences to suggest that though continents and oceans were formed very early in the geologic history of the Earth they have not remained stationary either in respect of their positions or forms. It found valuable support in the works of **Alexander du Toit** (1938); Arthur **Holmes** (1944); **Herman Hess** (1962-68); **F. Vine** and **D. Mathews** (1963); and **Tuzo Wilson** (1965). Contributions of all these scholars appear to have collectively found expression in almost concurrent works of **W.J. Morgan** of Princeton University, **D.P. Mackenzie** of Cambridge and **Bob Parker** of the USA in 1967. Their combined findings are believed to have laid the foundation of what came initially to be known as the New Global Tectonics. It was Morgan who is accredited to have proposed the outline of the hypothesis which later came to be known as Plate Tectonics. It is now more or less finally accepted that the Lithosphere which comprises continental masses as well as oceanshas been experiencing ruptures along weak zones somewhere along the mid-oceanic ridges.The plates, therefore, are moving away from each other in some sections of the Earth or are converging at some places. Gaps created by divergence of plates have been progressively

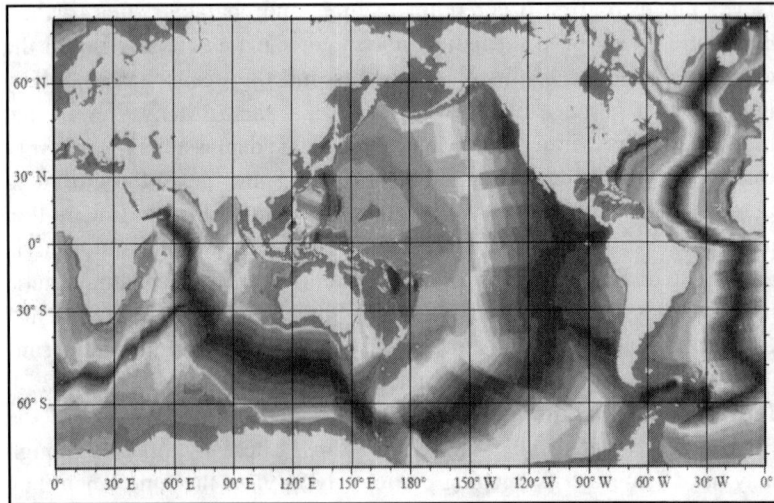

*Fig. 8.2: Sea floor Ages (from Arbogast: 2014:336)*

occupied by water masses. In areas of convergence seas appear to have been eliminated (*see module 4.3*). Analyses of rocks of most Oceanic bed suggest that most of them belong to Cretaceous Period that began some 135 mya. It implies that oceans and continents have not been permanent features as suggested by some scholars in early 20th century.. Ocean floors are found to be relatively younger than the cores of the continents and parts of the continents. Figure 8.3 gives an idea of age of different sections of the ocean floor.

## Reliefs of Second Order

The first order relief features are characterized by the presence of second order or what may be termed as intermediate level relief of land forms. They are found with some dissimilarity over continental blocks and ocean basins alike. Second order of relief on continents is represented by mountain chains, plateaus, lowland and plains. One may cite examples of the Himalayas, Alps, Rockies, Andes, Atlas, Appalachians, and Urals etc. as mountain chains. Peninsular India, Canadian Shield, Chinese and Russian Massifs as well as the Tibetan Highlands are representative continental plateaus. Other distinct second order reliefs on continents are extensive plains like Indo-Gangetic Brahmaputra Plain, Mississippi-Missouri Plain, The Nile Plain and many like them. Most deltas, marshes and swamps as well as extensive areas of inland drainage on the lands represent lowlands.

Oceans basins, similarly, possess many features many of them akin to continental second order features. They include features like continental rises, shelves, slopes, mid-oceanic ridges and deep sea plains. Yet there are some features which are found only in oceans like oceanic trenches and submarine canyons. The relief features which are found above the sea level for common understanding are called **'Positive Land forms'**. Below the sea level they are known as **'Negative Land forms'**. Many investigators have tried to explain the mechanism of development of these features. They generally have attempted to base their explanations on uniqueness of distribution of these second order reliefs believed mostly to be present since last 570 million years.

These second order features, however, are found to have some definite pattern of distribution both over land and below sea level as indicated below:

Most of the mountains are situated on the margins of the continents or erstwhile Shields (Fig. 8.2). It is more obvious in circum Pacific belt and Eurasia. It implies that the margins of the shields were characterized by the presence of narrow and mobile geosynclines which provided rooms for deposition of sediments obtained from sub aerial erosion of the shields. They ultimately became the sites of orogenic movements and evolution of mountains since Pre- Cambrian periods.

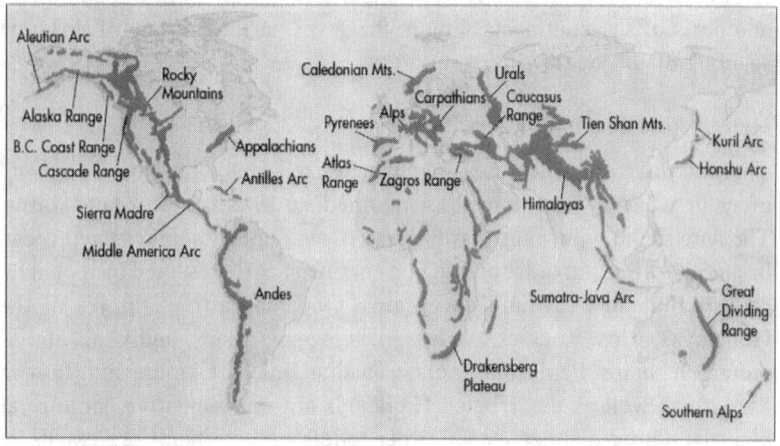

From Arbogast: 2014: 346.

*Fig. 8.3: Distribution of major continental mountain systems in the World*

Eastern coasts of the continents generally have broader extension of continental crust underwaterascontinentalshelves.ThePacific Ocean is the largest single feature on the Earth accounting for about 75% of the area covered by oceans.

All oceans with the exception of the Pacific and Arctic Oceans have almost centrally located mid-oceanic ridges. In the Pacific, the oceanic ridge is located in the eastern part of the basin off the coast of Central America to Gulf of California. Extension of the same ridge is found off the coast of north western USA till south western Canada. Arctic Ocean, on the other hand, is the only ocean which has two parallel ridges. All oceans have abyssal plains size and shape of which depends on the dimension of the oceans and location of the oceanic ridges. Western Pacific, Indian and Atlantic Oceans Have arcuate arrangement of active volcanoes which rise above the sea level. They are called **Island arcs**. Between the Island arcs and adjacent continents oceans are generally deeper which may vary in depth between 2 and 5 km. These sections of deep water along the concave side of island arcs are called **marginal seas**. These seas are mostly underlain by oceanic crust.

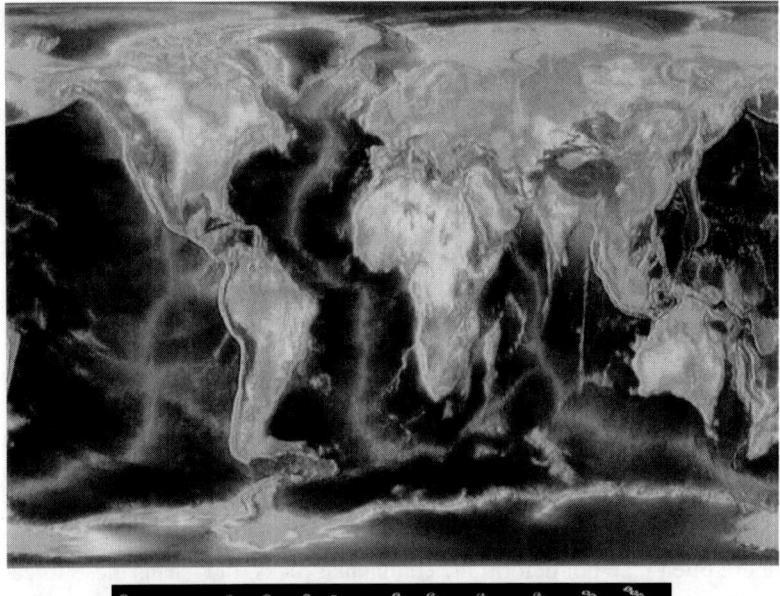

***Fig. 8.4: Distribution of Second Order Reliefs.***

On the convex side of all the island arcs is found maximum depth of water. They are known as submarine trenches. Similar trenches are also found along the continental margins having active volcanoes as is evident in case of Peru and Northern Chile. Mariana trench in the Pacific is the deepest of them all having a depth of 10,911 km. from the sea level. These trenches, however, are absent in areas where the volcanoes are very close to the land as in case of Alaska. It appears that the trenches in such cases are generally filled with new sediments. In areas where trenches exist no continental rise is found and even continental slopes are steeper.

## Reliefs of Third Order

Third order relief features represent the detailed configuration within the second order land forms. It includes details of feature at much smaller scales and which differentiate them from similar landforms elsewhere. The reliefs under this order may include identification of local landscapes in relation to elevation, ruggedness, and stability. Thus, a detailed relief of second order features may include features like hills, domes, valleys, canyons etc. Such details may, however, vary from one physico- climatic region to the other in their existing dispensations. It is the study and close analysis of them which may provide clue to the causes behind their evolution and earlier palaeogeographic environment.

## Earth Movements and Rock Deformation

Landforms both positive and negative are the results of a complex interaction of the forces operating within as well as over the surface of the Earth. The forces generated within the earth are called **internal** or **endogenic (endogenetic) forces**. Resultants of these forces are counter balanced by another set of **external** or **exogenetic forces**. These forces are responsible for sequential development of landforms and operate through denudation processes.

**Endogenic forces:** These forces are basically associated with the release of heat energy through upwelling of magmatic material from asthenosphere. These materials finding their way to the surface through volcanic vents are cooled and create the first order landforms. Similar subsequent activities but obstructed by the overlying continental crust more often than not deform the continental rocks. In Oceans these forces operate along the mid-oceanic ridges where new molten matters find openings along **axial rifts**[1] and create new surface. It causes movement of earlier solidified materials laterally on both sides of these ridges. It implies that the Earth's surface dimension must be enlarging. It does not happen. As a matter of fact, the same amount by which solid surface is created is consumed by down welling of matters in some parts of the Earth (for details see Mountain Building and Plate Tectonics). Thus, a circulatory motion of earth materials is established which forms and deforms the upper surface. These endogenic processes may be grouped in two categories- (i) *Volcanic*, and (ii) *Tectonic*. These forces set in a process of mutually complementary compression and tension in different parts of the Earth. Thus, a process of deformation of rocks or its layers is started which distinguishes one type of landform from the others. Volcanic processes are generally started when two crustal masses of different dispensation like oceanic surface and continental blocks collide with each other and the former goes under the latter. As a result the sea water along with eroded sediments from the adjoining land trapped at the margin of the oceanic crust is forced to go underneath. When critical temperature exceeds underneath combination of these matters with high proportions of silica in them is melted. The matters in this condition are known as *melt*. Melts generally have a very high proportion of Silica content (silica + almunium) normally exceeding 50% in their composition. They may be andesitic or silicic depending on proportion of silica in them. They are more viscous than normal magma. They move upward and reach the surface during violent volcanic activities to be solidified as extrusive rocks. Many times the silica rich viscous melts are arrested in subsurface where they cool slowly and are solidified over a long period of time as intrusive rocks. Volcanic eruptions find expression on surface mostly as **volcanic cones.** Intrusion of these melts as in case of batholiths, laccoliths leads to **up arching** of rock strata above and blockage of vents as in case of dykes.

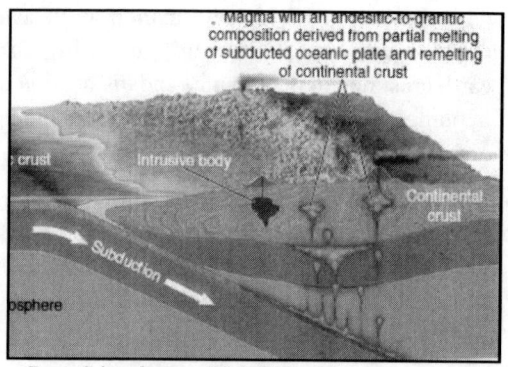

From Cristopherson (2012:337)
***Fig. 8.5: Rock deformation due to volcanic activity***

Tectonic processes, on the other hand, are the processes which cause bending or breaking of earth crust under the forces generated in the interior of the Earth. They are responsible for the creation of conditions of collision between similar or dissimilar separated crustal bodies leading to the formation of mountains and plateaus. Sometimes they may also lead to the formation of depressions or **down warping**. These processes are intrinsically linked with the denudation processes which are responsible for redistribution of pressure on underlying layers of the Earth and related processes. Many a time, thus, volcanic and tectonic activities together may be responsible for the development of landforms. And though the evolution of landforms generally has been associated with the operation of internal forces since the beginning of the Earth tectonic landforms are believed to have been formed mostly during the Quaternary earth movements. It is believed that over 90% of the present landforms, however, came into existence in post Cenozoic era most of them during quaternary period.

**Tectonic movements** may broadly be divided in two types (i) **Epeirogenetic** (epeirogenic) and (ii) **Orogenetic** (orogenic).

**Epeirogenetic movements** refer to vertical or radial motion of crust. These movements are responsible for uplift or subsidence of the crust or part of it in which rock strata forming them are though stressed they are disturbed or crumpled very little. It is a very slow process and operates over hundreds of millions years. Most of the table lands of the world with sharp scarps are formed in this manner. **Orogenetic movements,** on the other hand, are the horizontal or tangential motions of the portions of the crust. These movements may be *tensional* (extensional) or *compressional.* They are experienced in relatively weak or mobile zones of the Earth and are associated with deformation and shearing of rock strata. They are also known as **Mountain building forces.** Major inequalities in earth crust due to disturbance and dislocation of rock strata in the form of bending, folding, and breaking either due to epeirogenesis or orogenesis or both is known as *diastrophism.*

Tensional movements take place when either oceanic crusts are broken apart or continental crusts are fragmented. Thinning of crust allows fragmentation of its pieces which are pushed upward and flounder over denser materials below. *Block Mountains* and *Rift valleys* are formed due to these tensional forces affecting the rock strata along some faults.

Compressional movements squeeze the rock strata when lithospheric crusts collide with each other. It leads to deformation of rocks. Magnitude of deformation depends on the **stress** exerted and resistance offered by the rocks. Stress may be exerted on rocks by *tectonic forces, gravity* and *weight of overlying sediments.* Different rocks respond to stresses differently. These responses are essentially related to the elastic property of rocks. The way rocks respond to stresses is known as **strain.** Strain finds expression in rocks in the way they are deformed. Composition and nature of rocks make them either *brittle* or *ductile.* **Brittle rocks** refer to those rocks which break if their elasticity is surpassed. Most of the faults and shear fractures result from brittle deformation. **Ductile rocks,** on the other hand, are those rocks which are characterized by high elasticity and are not broken under normal stresses. Instead they are moulded in folds with different limbs and axes with alternate arrangement of *synclines* and *anticlines.* These properties of rocks are expressed in numerous types of faulting and folding. **Faults:** Faults are the fractures along which movement or displacement of rock layers takes place in the earth's crust. Most faults are the results of fractures caused by stresses in rock strata having brittle nature. These fractures are caused when the stress exceeds the strength of rocks. Stress in rock layers may be caused both by *tensional* as well as *compression forces.* These forces may displace parts of rock layers in any direction vertically as well as horizontally. The differential and proportional change in level of rock on two sides is known as *throw*. The side that plunges downwards is known as *down throw side*. The other side is identified as *up throw side*. Throws may range from a few centimetres to hundreds of metres. Fractures may also be created by lateral displacement which causes thinning and extension of area covered by rock strata. It is referred to as *heave*. These fractures are of two types - (a) *Tension fracture*, and (b) *Shear fracture*. Tension fractures are formed perpendicular to the plane of stress. Material movement on the two sides of fracture widens the fracture creating faults. Shear fractures unlike tension fractures are created within the fracture plane at an angle of around 30° to the plane of maximum stress where faults are created. Underneath flow of material in these types of fracture is generally parallel to the fracture surface displacing the rock layers. On the basis of these characteristics three types of faults may easily be recognized as- (a) *Normal faults*, (b) *Reverse faults* **(thrust faults)**, and (c) *Strike²slip fault* (see figure-8.6).

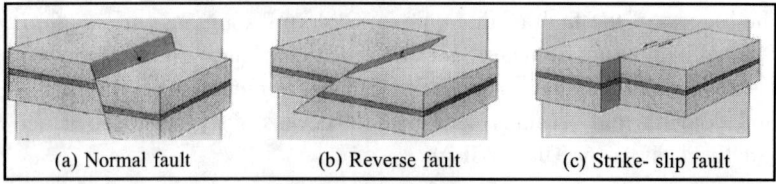

| (a) Normal fault | (b) Reverse fault | (c) Strike- slip fault |

*Fig. 8.6: Types of Faults*

**(a) Normal Faults**: Normal faults refer to the kind of faults which are created along vertical or inclined **fault planes**[3] in which down throw side is found along the **dip**[4] of the fault plane. These types of fault are created normally due to tension extending the faulted beds (*see fig. 8.6(1)*. Multiple normal faulting creates features like Horsts or Block Mountains#[5], Rift valleys or Grabens etc. One of the best examples of horst is provided by Shillong plateau in North East India. Bounded by sharp fault scarps the plateau is believed to have been separated from parts of erstwhile Gondwanaland and uplifted during the Himalayan orogeny of tertiary era. Conversely, rift valleys or grabens result from down sliding of crust strata between two or more normal faults. It is generally found in close proximity of horsts and is associated with horst formation. The Great Rift Valley of Eastern Africa and Red Sea are few examples of normal faulting with down sliding in between two faults. In India, River Narmada and River Tapi flow in faulted valleys.

**(b) Reverse Faults:** These types of faults result from compression of strata in which upthrow side is found opposite to the direction of dip apparently over riding the downthrow side. As a result faulted beds are shortened (*see fig. 8.6(b)*). It may be seen in areas where anticlines are broken and are pushed over the other strata of the fold. Most of the newly folded mountains of tertiary era are characterized by reverse faults consequent upon over folding (discussed in section on folding). Occasionally, however, rift valley like feature develops in reverse faults due to compression and wherein central block is forced down under lateral pressure. These kinds of valleys are known as Ramp Valleys. Compression forces exerted by the Himalayas in the north and Shillong Plateau in the south are believed to have forced down the River Brahmaputra to make it a Ramp Valley.

**(c) Strike- slip fault:** Different from above mentioned two types of faults strike- slip faults refer to the movement of rock strata sliding past each other in horizontal direction. They are created parallel to the strike of the fault plane without any scarp **(see fig.8.6(c)**. They are also known as shear faults, wrench faults, transcurrent faults and transform faults. The biggest of the fault of this type is 1200 kms. long *St. Andreas fault* of the USA running from Gulf of California to the North of San Francisco. It shows characteristics of a trench. There are numerous faults of this kind along the mid-oceanic ridges in all the oceans. They have been named transform faults by **Tuzo Wilson**.

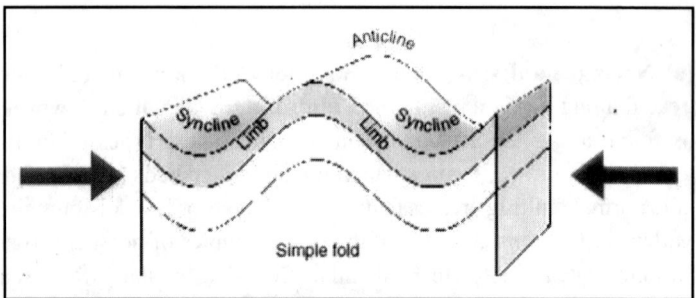

*Fig. 8.7: Components of simple folds*

## Folds

**Folds:** Folds are produced by compressive force in ductile rocks which essentially are sedimentary and horizontal in nature. Composed of layers of different strength if these rocks are exposed to tectonically generated compression they assume a shape having alternate arrangements of up arches and troughs. The line of maximum curvature of strata in folded rocks is known as *hinge*. The two sides of fold are known as *limbs*. Limbs meet each other at the axis of the fold and the surfaces that separate them are called the *axial plane*. Axes of up arches from where limbs slope down on opposite sides are known as *anticlines*. Similarly, the axes of the rock strata from where limbs slop up wards are known as *syncline*. Anticlines and synclines are inseparable and form an alternating bends in folded rock strata. Fig. 8.7 is a simplified representation of the integral components of folds.

Materials that make the rocks must have the flexibility to deform in tune with changing pressure and heat. Only sedimentary rocks have this kind of flexibility. It needs to be realized that the folding of rocks takes place under certain specific rocks conditions. Some important of them are listed below:

They must have plastic property to respond in tune with the increasing temperature. Rocks must be strong enough to bear the force of compression. If the force exceeds rocks internal strength they experience fracturing.

Shape and size of folding, however, depend basically on two factors- (a) *the magnitude and direction of compression*, and (b) *the nature of different strata forming the rock*. Folding is classified on the basis of inclination and orientation of their axes. It is the arrangement of limbs in respect of their axis that defines the types of folding. Folds having their axial plane vertical to the horizontal plane are known as *symmetrical folds* and arrangement of anticlines and synclines in them are at even distances. Though not very normal such folding are obtained when the compressive forces are either mild or equal from the two sides. If compressive forces are uneven axial plains tilt in the direction of lesser force and *asymmetrical folds* are created. Asymmetry of folds is proportional to the intensity of forces from both sides. Greater intensity of compressive force from one side makes the limbs inclined in the same direction. This may create *overturned* (one-limb vertical) folds. If the limbs of folds are found to be inclined in the same direction with parallel dipping beds they are known as *isoclinals*. Many times one limb of folds is found to be reclining over the other. These kinds of folds are named *recumbent folds*. Any increase in the intensity of folding may lead to fracturing of rock strata at the axis of recumbent folds causing forward displacement of the upper highly contorted limb. The upper limb may be placed some distances away from the *shear plane* from where it is

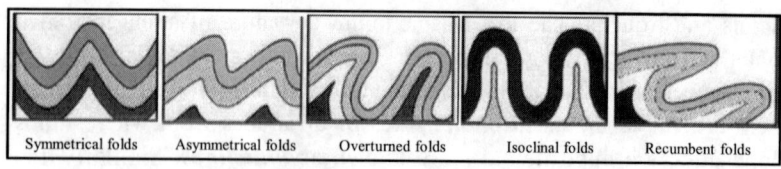

| Symmetrical folds | Asymmetrical folds | Overturned folds | Isoclinal folds | Recumbent folds |

*Fig. 8.8: Types of folds*

detached from the lower limb. They are called ***nappes***. The area between the upper limb and the lower limb of the rock strata is generally occupied by rocks of different origin. All these variants of asymmetrical folds are very common in newly formed fold mountains. Most common types of folds as discussed above are illustrated in fig. 8.7.

Besides above mentioned folds found generally in young rock strata some areas may have more complex and multiple developments of folds. Some areas having long history of structural deformation may be exhibiting many episodes of folding in which young ones are superimposed on the older ones. As fold usually is not found in isolation development of new fold system over an earlier folding create a very complex fold structure. It is found in some areas that numerous folds of recent origin are superimposed on earlier axial surface either of the same age or different era. If these folds develop over extensive anticlines they appear to form fan like structure. They are called ***anticlinorium***. Similar structures in vast areas of earlier synclines are called ***synclinorium*** (see figure 8.9 below).

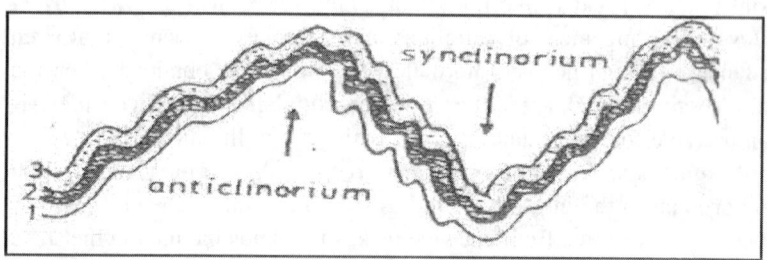

*Fig. 8.9: Anticlinorium and Synclinorium*

These kinds of features are very common in areas of shale deposits which are recrystellized to form slate in the processes of folding Central parts of the Pre Cambrian Aravalli ranges has the characteristics of a synclinorium bordered by anticlinorium on either side (Wadia, 1975:98).

One should, however, be careful about associating the apparent ridges to anticlines and valleys to synclines. Many times, owing to the differential nature of rock strata anticline may be eroded faster than the rock strata in syncline. It may lead anticlines to be lower than the syncline formations. It is very often reflected in the formation of synclinal ridge and anticline valleys. Following figure (Fig. 8.10) is an illustration of different facets of folding:

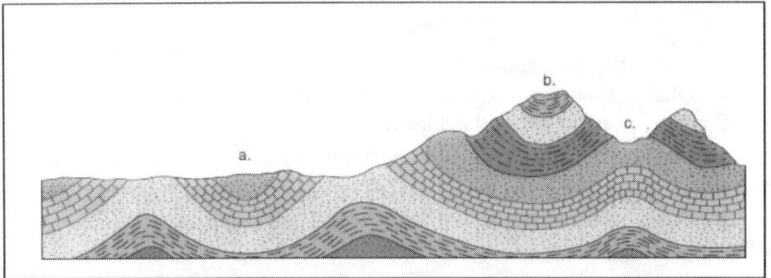

*Fig. 8.10: Facets of folding and topography*

It is obvious from the above figure that the bed rock structure and its configuration may not directly reflect in the topographic characteristics of a terrain which is the resultant of an intrinsically related exogenetic processes. It is more significant to know the nature of deformation of rock strata and exogenic processes operating over them and their sustainability when they are likely to be altered due to human intervention.

## Review Questions

1. What is meant by an order of relief? Give an example from each order.
2. Explain the difference between relief and topography.
3. What are the major differences between folding and faulting? What causes these types of crustal activity?
4. Explain compressional processes and folding, and *describe* four principal types of faults and their characteristic landforms.
5. How do intrusive processes and volcanism differ from each other?
6. Diagram a simple folded landscape in cross section, and identify the features created by the folded strata.
7. Define the following terms: igneous intrusion, pluton, batholith, and dike.
8. What is a **fault scarp**?
9. Describe and explain the formation of landforms that result from normal faulting (such as **grabens, horsts,** and **tilted fault-block mountains**).
10. There are basically two different forms of tectonic activity. These are _____. (a) compressional and extensional; (b) stressful and decompressional; (c) decompression and extensional; (d) compressional and stressful.

11. Compression leads to the folding of the crust, which results in the formation of ———————————. (a) anticlines and synclines; (b) synclines and troughs (c) upfolds and troughs; (d) troughs and anticlines.

12. _____ faults result in crustal shortening produced by compression of the crust. (a) Normal; (b) Transcurrent (c) Strike-slip; (d) Reverse.

13. What are marginal seas? How are they formed?

## Reference

1. Burchfiel, B.Clark, Robert J. Foster, Edward A. Keller, Wilton N. Melhorn, Douglas G. Brookins, Leigh W. Mintz, Harold V. Thurman (1982): Physical Geology; Charles E. Merrill Publishing Co.; London.

2. Dayal, P. (1990): A Text Book of Geomorphology; Shukla Book Depot; Patna.

3. McKnight, Tom L. and Hess D. (2011): McKnight's Physical Geography: *A Landscape Appreciation* (11th edi); Pearson Education, Inc.; Delhi.

4. Ollier, C. and Pain, C. (2000): *The Origin of Mountains;* Routledge.

5. Strahler, A.N. (2011): Introducing Physical Geography (5th edition); John Wiley & Sons; New Jersey.

6. Wadia, D.N. (1975): Geology of India (4th ed.); Tata-McGraw Hill Pub. Co. New Delhi.

## Notes

1. *Axial rifts denote to trench like formation at the highest longitudinal axis of the mid-oceanic ridges.*

2. Strike slip faults represents the direction of stratum perpendicular to the direction of Dip.

3. Fault plane refers to the surface along which faulting takes place.

4. Dip refers to the inclined stratum of rock in relation to the magnetic north in terms of angle and direction to the magnetic north.

5. Longitudinal and ridge like horst is called Block Mountain.

# Chapter - 9

# Mountain Building and Plate Tectonics

In Chapter 8 an attempt was made to provide an insight in the processes that impact the physical layout and characteristic features of the Earth. This chapter may be considered an extension of the previous chapter wherein an attempt has been made to investigate the changing understandings of mechanism leading to the formations of second order features above as well as below the sea level. Understanding of different viewpoints is considered to be important as these features are found to have multidimensional impact on physical as well as anthropological attributes in different parts of the world. This chapter, therefore, aims to undertake a study of the views on evolution of Orogenic belts and processes of mountain building emphasizing on the following:

Characteristics of orogenic belts;

Geosynclines and phases of Mountain Building;

Development of views on mountain building — Brief over view of earlier theories.

Plate Tectonics

## Orogenic Belts and Mountain Building

Mountain building in geological term is called **Orogenesis**. The term has been derived from the combination of two Greek words- 'Oros (mountain) + genesis (origin, beginning)'. All evidences indicate that mountain buildings are *episodal*. It implies existence of geological preconditions for mountain building where mountains are formed. Episode of mountain buildings is termed **orogeny**. These mountain building episodes are induced either *by migrating land blocks on collision course (accretion) and/* or *enlargement of continental margins* on account of sedimentation or *intrusion of plutonic granitic magmas*. These processes may operate independently or in unison. In any case, they result in thickening of the crust. Instability thus caused is believed to lead to deformation and uplifting of crustal rocks if strength of rocks is surpassed. Deformations and uplifting in these orogenes take place in the form of chains of folded or faulted mountain. Such areas have been identified as **orogenes**.

Orogenesis is a complex process operating over millions of year. Explanation of these processes involves understanding of geological history as well as ongoing denudation processes of the areas where mountains have been formed. In order to explain mountain forming processes one must understand the intrinsic characteristics of the orogenic belts. Following is an attempt to identify some major characteristics of Orogenesis and orogenic belts:

Mountain Buildings have been localized both in time and space. They have been spasmodic and cataclysmic followed by long periods of calm during which exogenetic forces sculpture the landforms (see chapter 4).

All mountains have larger linear extent than their width.

Mountains inevitably reflect crustal shortening.

All mountain chains are arcuate.

All mountain ranges as well as the 'island arcs' along the continental borders form 'Great Circles' though with different poles. The 'Great Circles' formed by them, thus, are found to intersect each other.

There exists a definite relationship between existing major and minor units of mountains and position of pre existing resistant 'stable blocks' of the continental masses.

Sedimentary rocks found in mountain ranges especially of the 'New Fold Mountains' show disturbance and deformation of horizontal strata on large scales.

Sedimentary rocks in mountain are found frequently to have fossils of shallow sea organisms. It suggests that mountain forming sediments were deposited in shallow seas.

Thickness of sediments suggests that the area of their deposition underwent subsidence.

Massive 'granitic intrusions' are common along the linear spread of mountains. It suggests rupture of the upper crust layers.

Based on above mentioned characteristics it may safely be assumed that the mountain buildings have taken place in areas where long and narrow but subsiding seas once existed. Irrefutable evidence was provided by **J. Hall** in 1859 following his study of sediments of Appalachian Mountains. He had concluded that those sediments were formed in a shallow and subsiding water body. It was **J.D. Dana** who in 1873 named such narrow troughs of water body **'geosynclines'**. As a matter of fact, scholars have increasingly realized that mountain buildings have intrinsically been related to different phases of evolution of geosynclines which generally were formed with great linear dimensions and much lesser width between two land masses. The most recent of such geosynclines has been identified as the **Tethys Sea** that existed since upper Palaeozoic through the Mesozoic era between the two great land masses of the time Eurasia ( part of the Angara land) in the north and the Gondwana land in the south. With a great linear extent between Gibralter in the west to the Indonesian islands in the east the Tethys is also suggested to have a width of about 3000 km. The space of the great geosynclines is presently occupied by the Alpo-Himalayan system of mountains. Present Mediterranean Sea is suggested by many scholars to be the remnant of the Tertiary Tethys Sea.

## Phases in Geosynclines' Evolution and Mountain Building

There is a general agreement amongst the palaeontologists that there are three major periods in the evolution of geosynclines. These periods are believed to gradually bring about changes that culminate in mountain building.

Period of lithogenesis,

Period of Orogenesis, and

Period of glyptogenesis.

**Period of lithogenesis:** The period of lithogenesis is characterized by accumulation of sediments in geosynclines leading to subsidence of their floors. It is believed to be affected in three ways: **(a)** Increase of sedimentary load on geosynclines floors and its consequent subsidence. Subsidence caused by the accumulating load is known as *sedimentation subsidence*. **(b)** Compression of the two sides of geosynclines due to subsidence caused by sedimentation that leads to further subsidence creating more room for sedimentation. It is known as *compression subsidence*. **(c)** Compressive forces working upon the accumulated sediments would also lead to folding of sediments slowly. Over folds and over thrusts would allow more sedimentation and further subsidence of geosynclines floor. It is known as *folding/ faulting subsidence*.

**Period of Orogenesis:** This period is believed to be impacted by operation of strong compressive forces and rupture of geosynclines' floors leading to strong folding and faulting. Faulting of the floor may also allow magmatic materials from lower spheres to intrude the overlying layers of sediments. All these processes together uplift the sediments proportional to the intensity of force during the period.

**Period of glyptogenesis:** This period basically refers to the time span during which isostatic equilibrium is attained and the characteristic surface forms are sculptured by erosion.

Despite broad consensus about these phases there have been a wide range of opinions as to the mechanism that could cause mountain building. All those opinions appear to have led to refinement of successive views. They, therefore, are worth of review. It is in this light that an attempt has been made below to assess some important explanations made earlier and their contribution to the development of presently the most accepted theory of Plate Tectonics.

Some scholars believing in the permanency of continents and ocean basins invoked thermal loss from and consequent contraction of the planet Earth to explain mountain building. Others subscribed to the view that the mountain buildings have been the results of displacement of continents during their long geological history. Mountains, according to them, have gradually evolved to attain their existing forms and shape. The views on mountain building, thus, may broadly be categorized under two broad schools of thought namely (1) **fixism**, and (2) **mobilism**. The orthodox view of fixism delved upon a permanent configuration of continents and ocean basins. It found support through the studies of **T.C. *Chamberlin, E. Suess, L. Kober, J. Joly and H. Jaffereys.*** Views on mobilism, on the other hand, had been finding greater support since 1924 when the 1912 study of *Alfred Wegener* became popular after its English translation. Echo of mobilism though is found in Argand's views on formation of Appalachian Mountains, in post First World War period particularly after 1926 studies on mechanism of mountain building by ***R.A. Daly,*** and ***Arthur Holmes*** involves disruption of earlier configuration of the Earth Crust through mobility. The view has found more support recently from the protagonists of **'Plate Tectonics'** backed by the studies of *Herman Hess* (Sea floor spreading) , *Tuzo Wilson* (Transform faults), *F.J. Vine and D.H. Mathews* (Magnetic anomaly Patterns).

## The Geosynclinal- Orogen Hypothesis of L. Kober

**L. Kober**, a German geologist like his fore runners **Suess** and **Argand** was basically a Contractionist. He like many others of his time believed that the Earth has been experiencing contraction since very beginning. It has been its cooling that initially created rigid masses or table lands of the Earth. These parcels of land were surrounded by more mobile zones of water. He called the primordial land blocks **Kratogens** and the shallow mobile body of water (geosynclines surrounding them) **Orogens.** Kober concurred with the view that there had been six major mountain building periods in the known geological history of the Earth. All these mountain building periods underwent more or less similar sequence of events. In explaining the sequence of events Kober was able to explain many of the characteristics of the existing mountain ranges. He had also been able to explain the addition of land area to earliest 'core lands' (kratogens) much in tune with the studies of Suess and Argand as well as southward shifting of orogenic belts during geologic times. He outlined the evolution of mountains as follows:-

The rigid masses (core lands or kratogens) were separated by long and wide geosynclines. The rigid masses on the two sides of the geosynclines formed the **'forelands'**'*. Sediments from the forelands were deposited in the intervening geosynclines.

The load of sediments on Geosynclinal floors would further depress their floors. It would make room for more sedimentation.

Depression of the Geosynclinal floor would bring the two forelands (kratogens) closer. Coming closer of the two forelands would lead to folding and over thrusting of sediments along the margins of the two forelands. It would form 'Border ranges' termed by Kober as Randkatten.

If compressive forces would be intense the folding/ over thrusting of the two Border Ranges would over lap each other and the entire sediment in the Geosynclinal basin would be squeezed. The two 'Border Ranges' can be identified along a line known as 'Narbe'.

In case of less severe compressive force, however, the part between the two Border Ranges would be little affected and stand distinctly between them. This unaffected land between the two Border Ranges was called 'Median Mass' by Kober.

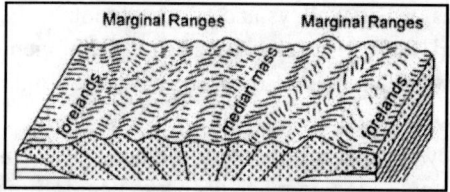

**Fig. 9.1: Hypothetical representation of Kober's View**

These mountain building movements would be characterised by deep seated tectonics and a high degree of metamorphism. It is so found in all areas affected by orogenic movements.

## Merits of the Hypothesis

Kober's view was considered credible on many grounds as it could explain many features that were considered difficult to explain earlier. Geosynclinal Orogen Hypothesis could explain -

Larger linear extent and lesser width of the mountains.

The great thickness of sedimentary rocks in Fold Mountains and associated fossils.

Compressive features.

Expansion of land area along the core land due to addition through mountains on the margins e.g. Eurasia and parts of Gondwana land-Peninsular India may be said to have been joined together by development of mountain ranges like Kunlun, Tienshan and Elburz with (i) Tibetan plateau as median mass between the Kunlun and Himalayan mountains; and (ii) The Anatolian plateau between the Elburz and the Iranian ranges of mountains. Similarly, Europe may be suggested to have expanded between the Fenno- Scandinavian and Russian plateau during successive Caledonian, Hercynian and Alpine mountain building periods. Kober identified many median masses in Alpine section namely (i) the Adriatic Sea (Foundered land mass) between Dinaric Alps and Apennines; (ii) the Hungarian plain between the Carpathians and Dinaric ranges; (iii) the Islands of Corsica and Sardinia (parts of the submerged median mass) between Apennines and Baltic Cordillera. In North America one of the median mass is represented by the Basin Range between Sierra Nevada and Wasatch ranges of mountains.

Arcuate form of the mountain ranges.

Granitic intrusion common along the linear extent of mountain ranges.

Periodicity of mountain building revolutions.

Occurrences of Rift Valleys and Block Mountains may be attributed to tensional forces initiated by crustal (Kratogenic) movements. Crustal movements may also cause superficial folding like those of Jura Mountains.

Kober's view is also in tune with the principle of isostasy wherein folding is supposed to go much deeper in denser Sima than above the Crust. Thus, even with the end of orogenetic revolution uplift of mountains would continue. This also is in conformity with the hypothesis of alternate subsidence and upliftment in the orogenic belts of the World.

## Demerits of the Hypothesis

The compressive forces generated by Contraction and movement of 'forelands' towards each other are considered insufficient to produce huge mountain chains many of which might have formed simultaneously.

The hypothesis has been able to explain mountain building where at least two rigid masses of lithospheres' crust are believed to have existed as in case of the Caledonian and Hercynian mountain systems in North America and Europe (along Candian Shield and Scandinavian Table land in the west) and the mountain ranges between Fenno-Scandinavian core land and Russian Massif in the east or more recently Alpo-Himalayan system of mountains between Eurasia in the north and African- Indian Peninsular land mass in the south. However, the hypothesis fails to explain the formation of North-South trending mountains like the Rockies and the Andes which are bounded on one side not by any rigid land mass but by the massive and deep water body of the Pacific.

Despite its short comings the Geosynclinal-Orogen hypothesis is still considered to have laid foundations to many later hypotheses including those under expansionist viewpoints e.g. of **Argand** (1924) after the publication of Wagener's 'The Origin of Continents and Oceans'. He appeared to have accommodated before **Kober** the view of coming closer of slower moving foreland and faster moving back land or hinterland (Kober's forelands).

## Radioactivity and History of the Earth: J. Joly

**Joly**'s hypothesis of mountain building was based on very simple and now universally accepted concepts. His views were based basically on the following assumptions:

The continental blocks are composed of lighter materials as compared to those of the ocean floors. In other words, the upper crust composed of sial (silica+ alumina with predominance of granitic rocks) rested over denser sima (silica+magnesium) of the oceanic floor composed predominantly of basaltic rocks;

All rocks are more or less radioactive and generate heat after their disintegration but the sialic rocks are found to be more radioactive hence more heat generating than the sima rocks;

A fare amount of heat from sialic crust (due to its exposure to atmosphere) and upper sima (due to its contact with water) is lost;

The lower layers of sima under oceans and its upper layers under the continental blocks would be experiencing a condition where heat would be accumulated.

Based on his assumptions and on the basis of thermal gradient Joly concluded that the average thickness of the sialic blocks was about 30 Km (present scholars suggest the average thickness of the same to be 35-36 Km) with a temperature of about 1050°c. Joly calculated that a thermal condition leading to melting of the basaltic substratum (1150°c i.e. 100° more than the obtained condition) would be reached between 32 million years and 56 million years. During this period he visualized a chain of events to recur leading to mountain building as follows:

The substratum (sima layer) undergoes liquidification due to increase in temperature.

Density of substratum decreases due to liquidification that leads to submergence of continental blocks in it.

Seas transgress the continental margins and Geosynclines are formed.

In these geosynclines sediments from land are deposited (the stage of lithogenesis)

Liquidification of basaltic substratum would also lead to increase in the effects of tidal forces. It is suggested to have affected westward movement of the continental blocks.

It would expose the oceanic floor of sima which would lose its temperature through conduction. It starts the process of resolidification and rocks attain their earlier density.

The continents are pushed up and gradually attain their previous height.

Buoyancy of continents (capacity of continents to float over the substratum) also decreases.

Transgression seas simultaneously start withdrawing (period of regression) and expose sediments on margins of continents to pressure caused by shrinking radius of the earth due to solidification of stratum (a zone where maximum expansion takes place due to melting and contraction caused by resolidification). Lateral compression, thus, created on sediments lead to their buckling and folding (period of glyptogenesis).

Complete resolidification of substratum causing uplift of the continental blocks may also be considered to uplift the mountains in second stage of its formation.

In this way, both the processes that of (i) lateral compression leading to warping and folding of sediments and (ii) vertical uplift necessary for isostatic equilibrium of the mountains are explained to a great extent by Joly.

## Merits of the Hypothesis

Transgression and regression of seas being common phenomena the hypothesis is able to explain sedimentation of the water basins and folding of sediments.

The hypothesis is able to explain the periodicity of mountain building and conforms to the known mountain building episodes.

It meets the requirement of the cyclic order of mountain building.

Suggestion that bigger the water body greater pressure would be exerted on sediments to form proportionate mountains finds supports in coastal Circum Pacific ranges of mountains.

Despite remarkable merits the hypothesis, however, is found to have over looked some very glaring facts and which cannot be explained under assumed mechanism. Some of facts that make the hypothesis untenable are as follows:

The hypothesis fails to explain absence of new fold mountain ranges along the Atlantic border.

It fails to explain the formation of mid-continental Alpo-Himalayan mountain systems.

**Joly** visualized Orogenesis to take place during fixed periods of 32 million to 56 million years. It has not always been found to be so.

**Jaffereys** pointed out that the magnitude of tidal forces invoked by **Joly** for westward movement of continents was inadequate.

**Jaffereys** based on his calculation considered it highly improbable that recurrent melting and solidification of basaltic substratum would take place as assumed under the hypothesis.

According to the postulates of the hypothesis mountain should be formed all around the continents. It is not found to be so.

Transgressing seas is supposed to form geosynclines wherein deposited sediments are moulded to form mountains but many studies have revealed that sediment yields there in may not be sufficient to explain formation of high mountain ranges.

These criticisms though have made Joly's view redundant its significance lies in the fact that it paved the way for refinement of later more acceptable views of Holmes.

## Thermal Contraction Hypothesis of Harold Jaffereys

Thermal Contraction Hypothesis represents one of the other Contractionist view point proposed by **Jaffereys** in 1929. The hypothesis is based on a simple comprehensible idea of accommodation of cooled and solidified lithosphere to still shrinking inner layers in which it was suggested that no significant thermal change could take place below a depth of 700 Km. As a result, no appreciable contraction or volumetric changes could take place in layers below it. Jaffereys, therefore, restricted the mechanism of mountain building to layers above it. He formulated his view with the help of complex mathematical reasoning in which he also took note of rotational velocity of the Earth. He out lined the formation of mountains as follows:-

- Different layers/shells of the Earth above 700 Km. have been cooling differentially to attain solidification.
- It led to differential contraction and volumetric changes in solidified layers in which each layer would cool and contract more than the layer just below it.
- Contraction of the upper layer which could not contract any further led to reduction in its radius.
- But reduction in radius would be obstructed by the relatively warmer layer underneath it.
- In course of time, as a result, the outer layer became too large to fit on the inner shell.
- It led to collapse of outer layer on to the inner contracting regime (it is believed that the portion where the greatest cooling would be taking place would also become too small to fit on the inner shell. It implied that there had to be an intermediate "level of No Strain" where the upper contracting layer fits to the inner layer).
- Below this "level of No Strain" any contraction would cause reduction in the volume of the layer causing down ward shift of the same and making the cooling layer too small to fit on the interior.

- In order to allow the upper shell to fit, the interior shells must be stretched horizontally implying lateral spreading and thinning out of their materials.

- This thinning would form fissures filled by materials underneath.

- Above the 'level of no strain', horizontal compressive stress resulting from a decrease in the radius of the upper layer would lead to buckling and folding, hence mountain formation.

It is suggested that the range of cooling would be greater beneath the oceans where rocks are generally basic and stronger than under the continental blocks which are characterized by acidic and lighter rocks.

These characteristics are believed to have caused compressive forces around the ocean to act away from it towards the land. **Jafferys** believed that the Circum-Pacific mountain ranges were formed due to such compression.

Additionally, uplift an essential corollary of mountain building would ensue following the collapse of upper layer on to the interior and consequently its melting and subsequent expansion.

Though the views of Jaffereys are not much appreciated now his hypothesis do have some merits particularly in view of the time when it was proposed.

## Merits of the Hypothesis

Jaffereys invoked contraction of the Earth partly through its cooling and partly through reduction in the speed of its rotation (both are accepted by all the scholars).

The hypothesis claimed a crustal shortening of a magnitude of about 200 Km. and consequent reduction in surface area of about $5 \times 10^{16}$ cm$^2$. Such shortening of crust is in conformity with the present knowledge and measurements.

The hypothesis is also able to explain the distribution of mountains, igneous intrusions and periodicity of mountain building without invoking any kind of continental drift as invoked by Wegener

Thermal Contraction hypothesis of Jaffereys, however, has been found lacking in many of its explanation. Some of them are pointed out below:-

Cooling and contraction are considered a continuous process. It, therefore, is believed to lead to minor folds and not major mountain ranges spread in different nook and corners of the Earth.

On the basis of this hypothesis the interval between two mountain building periods should go on increasing. In reality it is not found so. In fact, most of the mountains of the tertiary era were formed within 200 million years of previous Hercynian orogeny. In such a short period contraction of the magnitude suggested appears improbable. Jaffereys though provides a quite reasonable account for the formations of coastal mountains he appears to have failed in providing adequate explanation for formations of much bigger mid-continental mountain ranges.

In the light of the above mentioned observations by different scholars Jaffereys' Thermal Contraction hypothesis has now been rejected as a plausible explanation of mountain building mechanism.

## Continental Drift Theory and Mountain Building: Alfred Wegener

**Wegener**, a German meteorology professional with basic training in astronomy put forward his views in the form of a theory on migrating continents in 1912 (see chapter 4). He fully elaborated the idea in his book **'The Origin of Continents and Oceans'** published first in 1915. Wegener went on adding additional inputs to his theory in his subsequent revision of the book in 1920, 1922 and 1929 respectively. In the process he also tried to explain distribution of mountains and their formations.

It may be noted that since late 16th century many scholars had started believing that the continental blocks had not been stationary. It appears that the Dutch cartographer **Abraham Ortelius** in 1596 was the first to suggest disintegration and separation of Americas from Europe and Africa. Even **Francis Bacon,** the British Philosopher inferred in 1620 some kind of continental movement when he noticed the likeness of continental margins on two sides of the Atlantic. In 1780 **Benjamin Franklin** based on findings of fossilized oyster shells in British uplands also suggested some kind of crustal movement. **Alexander Von Humboldt** during his long voyages across the Atlantic also realized the similarity of its eastern and western coasts. Later, **Antonio Snider- Pellegrini**, a French geologist, in order to justify the findings of similar kind of carboniferous fossils and coal seams both in Europe and North America, gave an illustration of closed and opened Atlantic Sea in 1858. Similarly, **F.B. Taylor**, an American geologist in order to explain the peculiar pattern of distribution of tertiary fold mountains and which contraction theories failed to account for, resorted in 1910 to some kind of crustal displacement. He attributed the north and westward crustal movement of the Cretaceous' Lauratia (then located near the North Pole) and Gondwanaland (near the South Pole) to the tidal force of the moon. It may be noted that when Wegener independently published his account of continental drift it was found to be so close to that of Taylor's that the theory for some time came to be known as '**Taylor- Wegener hypotheses**. Wegener himself acknowledged the contribution made to his theory by Taylor besides many others from mid 19th century. Wegener based his study of continental drift on evidences collected from a plethora of sources and from his own discoveries. In the process he made certain assumptions as listed below:

- He believed that there existed a super continent and a super ocean some 250 million years ago during Carboniferous period named respectively as Pangaea and Panthalassa.

- He presumed that continental blocks comprised of lighter sial and deep seafloors were composed of denser sima. He also assumed that the continents rested over denser sima ocean floors and extended downwards.

- He assumed that the difference between the centre of gravity and centre of buoyancy of an ellipsoidal Earth (believed to be acute around 450 latitudes) caused disintegration of the continental mass and northward movement of their parts from around southern Pole.

- The tidal forces of the Sun and the Moon being stronger during this period dragged the disintegrated continental blocks toward the west.

- The North and South Poles were located in the Pacific and Natal of South Africa respectively *(changing positions of Poles were supported by Vine and Matthews.*

In order to explain the formation and distribution of mountains in the world Wegener suggested that:

➤ Pangaea composed of light sialic matters floated over denser sima floor of the Panthalassa.

➤ Two tangential forces, (i) gravity of the Earth and (ii) tidal force of the Sun and Moon, operated simultaneously to pull Pangaea apart.

➤ Gravity was assumed to have induced its equator ward movement and the tidal force dragged it towards the west.

➤ In course of time, Pangaea disintegrated along the then equator and formed two distinct land masses of Angara land (comprising of ancestral parts of North America, Greenland, Eurasia, and Tarim Basin) and Gondwana land (consisting of cores of South America, Africa, Arabia, Persia, Peninsular India, China and Australia). These two land masses were separated by a Geosynclinal depression running from east to west. The Geosynclinal depression has been named as Tethys Sea (*see chapter-4 and fig.4:3*).

➤ Further northward movement of the Gondwana land led to separation of Africa, Peninsular India, China and Western Australia from the Antarctica.

➤ The north ward movement of Africa and Peninsular India created the Indian Ocean whereas west ward drift of the Americas formed the Atlantic.

➤ Tethys Sea experienced sedimentation from Angara land in the north and Gondwana land in the south. In the meantime, northward movement of Africa and Peninsular India accentuated by deepening bed of the Tethys led to squeezing and folding of sediments. Thus evolved the tertiary folded Alpo-Himalayan mountain systems of Eurasia and the Atlas of north-western Africa.

➤ The west ward drift of continents under tidal force and the resistance posed by rigid sima layer of the Pacific caused formation of the north-south trending western cordilleras of North and South America respectively.

## Merits of the Hypothesis

In his attempt to explain the terrestrial attributes of continental blocks and ocean basins Wegener provided a number of evidences from different fields of paleontology, paleoclimatology and paleobiogeography. They have withstood the test of time though they were not accepted initially. It may be attributed to lack of scientific information then. Below are some evidences which did not find much support earlier but after the end of the 2nd World War many scholars appear to be convinced about those evidences. Few of those evidences with later commentaries are given below:

- Resemblance of coastlines on the western and eastern sides of the Atlantic which appear to fit like Jigsaw. In fact, it was found by **Bullard, Everett et al**. through computer fitting in 1965 that the South American and African coastline could fit well at a depth of 500 fathoms (about 1000 meters) below sea level (same type of fittings could be found by **Sproll** and **Dietz** for Australia and Antarctica in 1969 and by **Smith** and **Hallam** for South America, Africa, Arabia and Antarctica in 1970).

- Remarkable correspondence of structural attributes of eastern North America and Western Europe e.g. the Appalachian in eastern U.S.A. may be extended across the Atlantic to the Hercynian folds of Ireland, Wales and central Europe. Same may be said of the Caledonian folds in North America having a south west trend and re-emerging in the same direction in Ireland and Scotland with remarkable structural unity. *It was supported later by South African* geologist **Alfred Tuzo Wilson** (1937) who found close geological similarity along the African and Brazilian coastal regions.

- Availability of coal deposits in humid middle and high latitudes is suggestive of the areas having experienced wet tropical climate in their geological past.

- Occurrences of glacial deposits of Permo-Carboniferous periods (when super continent existed) over tracts of Brazil and Peninsular India. No trace of such glaciations is found over the parts which made Eurasia (then believed to have tropical conditions). It suggests that there had been extensive glaciations nearer the southern pole. Similarly, occurrence of sandstone, rock salt windblown sand and gypsum presently in entirely different environment is suggestive of continental wandering.

- Similar fossils of terrestrial and oceanic animals over far flung areas and widely separated land blocks are also indicative of contiguity of lands in the form of super continental blocks.

## Demerits of the Hypothesis

In spite of the fact that the idea of continental drift has found a lot of support in recent times, Wegener was severely ridiculed by his contemporaries particularly from the American Universities. Most of the criticisms were directed towards Wegener because he was considered to have transgressed the specialized domain of geology. The basic flaws in Wegener's hypothesis, however, have been with regard to the forces invoked by him to induce continental drift and mountain building. They may be elaborated as below:

■ The forces invoked by Wegener the gravitational and tidal forces, are not sufficient to move the continental masses. Present quantum of gravity is not found to be sufficient to move the continents northward. It was also calculated that in order to induce westward movement about ten thousand million times greater tidal force than that of the present would be required. Tidal force of such a magnitude, however, would have stopped the rotation of the Earth.

■ The formation of the western mountain systems of North and South America as suggested by Wegener is found untenable. It was suggested that continental masses were floating over sima. Thus, there could not be much resistance offered by the oceanic floor to lead to the formation of Western Cordillera.

It may however, be noted that other criticism of Wegener with regard to the jigsaw fit of the continents and geologic similarities of the coasts have been increasingly finding support. Arthur Holmes, a British geologist, even attempted to provide an alternative mechanism for continental drift and mountain building in 1928.

## Hypothesis of Sliding Continents: R.A.Daly

**Daly** in his book 'The Mobile Earth' published in 1926 suggested the mobility of the continents to have resulted in mountain formations. His suggested mechanism of mountain formation is based on his assumptions that

• There existed three primordial land masses close to the two poles and one between them. They comprised nine table lands separated from each other by narrow and shallow water bodies identifiable with geosynclines and the primeval Pacific Ocean.

• The land masses were composed of denser rocks and the substratum of relatively less dense rocks having proportionate radioactive components,

• Lands would be sloping towards the geosynclines and the early Pacific Ocean controlled by the Earth's gravity.

Based on these assumptions he, then, proceeded to outline the evolution of mountains as follows:-

➤ The Earth crust being a bad conductor of heat cooled fast (due to its exposure to the atmosphere) and contracted to its limit.

➤ The interior, however, would continue to lose heat slowly to the exterior (due to obstruction) but leading to its contraction away from the outer shell.

➤ This would necessitate the outer shell to gradually collapse and not fall indiscriminately in order to fit over the cooling interior shell.

➤ This collapse has been suggested to have happened in two ways:-

(a) Under the weight of the oceanic water, and

(b) Under the weight of the Geosynclinal sediments.

➤ Corroded materials of the table lands would be carried towards the primeval Pacific and inter-continental geosynclines.

➤ Down ward pressure of sediments on the Geosynclinal floors would exert lateral pressure on higher continental domes relieving pressure on materials underneath them and causing their out ward expansion from the centre of the Earth hence making them lighter.

➤ In order to attain balance, therefore, matters from 'sediment compressed region' would flow towards the domes.

➤ It would make domes to have excess of matters leading to rise in their elevation causing more erosion and sedimentation as well as strain on the geosynclines floor.

➤ Excess strain would lead to rupture of the geosynclines floor forcing it to slip under the glassy basaltic sima below and large continental blocks to slide towards geosynclines.

➤ The marginal sediments would be much squeezed and folded heralding the first phase of orogeny during which period denser substratum materials would come below not allowing them to sink.

➤ Lateral pressure caused by sliding continental blocks over lower sediments would push them down ward to be increasingly

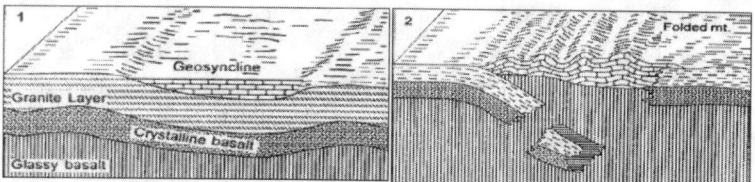

*Fig. 9:2: Illustration of Sliding Continents*

heated. It would cause melting and expansion in their volume leading to the progress of second phase of orogeny when mountain ranges are uplifted.

## Merits of the Hypothesis

Above mentioned processes suggested in the hypothesis has been found to explain many of the existing features of the mountain configuration in the world and may be summarized as follows:-

-   Formation of the Alpo-Himalayan system of mountains is attributed to sliding of continental blocks of Laurantia and Gondwana land towards the mid-latitude geosynclines now universally accepted as the Palaeozoic - Mesozoic Tethys Sea.
-   Formation of the north-south trending Rockies and Andes mountain systems is also explained on the basis of American continental dome's slide towards the Pacific depression.
-   Island and mountainous arc of eastern Asia is also accounted for by the Pacific ward slide of the Asian block.
-   Existence of the trenches in the Pacific according to this hypothesis may be attributed to the downward pressure exerted by the Island arcs.
-   It also explains why mountains are found only along the Pacific coast and not along other oceans.
-   Daly attributed the evolution of other oceans to tensional rifts a necessary corollary of mountain forming compressive forces.

## Demerits of the Hypothesis

Daly's views, however, have been found by many later scholars to be based on a number of erroneous assumptions. They, therefore, suggest it to be untenable. Basic demerits of the hypothesis are outlined below:

-   Daly simply presumes the distribution of earlier continental blocks, mid continental furrows and the primordial Pacific Ocean. He failed to give any cause for such distribution.
-   He also failed to explain the existence of mid-continental fold mountains.
-   His assumption that substratum is less dense than the continental blocks is erroneous.

Despite its rejection, the hypothesis has been found to provide solid foundation for development of later hypotheses particularly those of

Holmes and later more integrating theory of the Plate Tectonics. In fact, his contribution, **Letsch** (2015) believes, was the turning point in converting the American view point in favour of moving continental blocks and which certainly influenced the development of later theories.

## Convection Current Theory: Arthur Holmes

In view of the remarkable evidences collected by Wegener in favour of continental drift but his failure in providing a logical mechanism for the same **Holmes** proposed his theory of convection current in 1928 (*John Perry had already visualized in 1895 the possibility of convection currents in the interior of the Earth*). Being a geologist himself he based his propositions on geological facts available at the time. Like Joly and Daly he also invoked radioactivity of the rocks to generate differential heat conditions within different layers of the crust and upper sub-stratum (he believed that almost all the radioactive energy produced by the rocks is converted into heat). It also implied that greater the thickness of the crust greater would be the disintegration of radioactive minerals hence greater production of heat energy. In the light of the foregoing he made certain assumptions to explain continental drift and mountain building:

- Total loss of heat from the surface of the Earth to the space would be equivalent of the heat generated by 60 km thick crustal shell. On the basis of the heat generation capacity he divided the crust in three broad layers-

    ✔ Upper granitic layer with most dominant radioactivity (confined to continental blocks),

    ✔ Intermediate layer of mixed rocks with diminishing radioactivity (mostly oceanic layers with the exception of continental islands), and

    ✔ A crystalline lower layer above the sub-stratum (with still lower quantity of radioactive rocks).

- These layers would have uniform depth below the continental and oceanic crust.

- Irrespective of very low heat producing capacity of the sub-stratum the crustal layers impede the loss of heat in proportion to their thickness i.e. the continental masses block the heat more than the oceanic crusts.

- It, therefore, would generate a very strong convective current in the sub-stratum under the continents and a weak one below the oceanic crusts.

- The under belly of continental blocks where ascending convective currents would be diverted becomes an *area of constant tension.* On the other hand, the zones where the sub continental and sub-oceanic currents converge would become a zone of *constant compression.*

- Tensional and compressive forces continue to be operating till configuration of different crustal layers is changed.

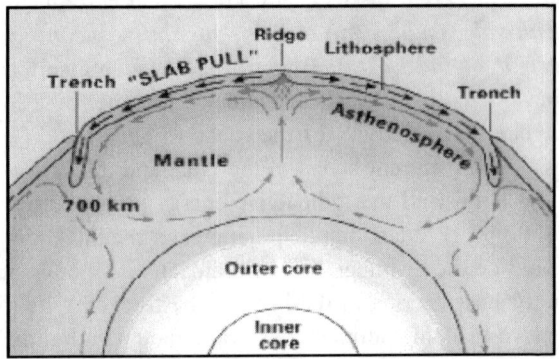

*Fig. 9:3: Convection currents after Holmes*

In the light of the above assumptions **Holmes** outlined the changes including mountain building that the Earth crust undergoes as follows:

➤ Despite low heat generation capacity heat is accumulated in the sub-stratum below the crystalline glassy layer of the crust.

➤ Consequent upon temperature gradient heat is accumulated more under the continental crust and less under the oceanic crusts.

➤ Due to this differential accumulation of heat strong ascending currents are generated under the continents and a weak convective current under the oceanic crust and their margins.

➤ Similarly, the ellepticity of the Earth (limited only to the outer shell and which leads to greater crustal thickness along the equator) causes generation of stronger ascending currents and their diversion underneath making it a zone of tension. Due to lesser temperature gradient in the Polar Regions the currents descend down to the substratum to complete the circuit of the convection current. Such descending currents would also be found everywhere under continental shelves where currents from underneath the continents and oceanic crusts converge.

➤ Diverging currents under the continents would lead to its thinning and rupture as well as drift of its parts away from each other.

➤ It would cause formation of new but shallow ocean basin (geosynclines) and allow the excess of heat to escape.

➤ Where the outward currents from beneath the continents meet the opposing currents from under the Oceanic crust a zone of strong compression would develop. It would force them to sink towards higher thermal layers in the interior of the Earth.

➤ It would lead to density change of materials due to their recrystellization and phase change. This change would also lead to acceleration of descending currents.

➤ At the same time, the outward moving current under the continents would lead to thickening of sial margins inducing the first phase in mountain building. (Geosynclines which are suggested to be forming along the margins of the continents where sub-continental and suboceanic currents converge are also considered to be the areas where sedimentation leads to hastening of descending currents. Simalteneously, the lateral compression from the two currents would lead to folding of the sediments in the geosynclines with increasing compression. It would bring about continuity in the forces of mountain building).

➤ As the mountain roots having less dense materials than those of the sub-stratum would not be able to sink into it would melt due to high temperature there in and fuse with them. It would produce igneous activities.

➤ With the formation of new ocean basins with the rupture and drifting of continents as well as the phase changes of rock materials the convective currents gradually would die away and give rise to new sets of convective current. Thus, conditions for next mountain building episode are induced.

## Merits of the Theory

**Holmes** seemed to have logically explained most of the objections faced by his predecessors and which have been integrated in presently more acceptable theory of the Plate Tectonics. He also gave plausible explanation for the present distribution of major landforms of the Earth.

- He has satisfactorily explained how the primordial continent of Pangaea ruptured and broken to form the Tethys Sea.

- He could also explain the formation of the Atlantic Ocean.
- The theory also explains the formation of mid-continental as well as coastal mountain ranges.
- Igneous activity associated with mountain building is explained, and Periodicity of mountain building is explained.

## Demerits of the Theory

Due to lack of hard evidences at the time when the theory was proposed Holmes faced criticism mostly in respect of

- Existence of convective currents in the interior of the Earth (Presently, however, most of the Earth scientists believe that convection currents do exist and immensely influence surface features).

- The ability of the convection currents to carry continental apart has been a major question against the theory (*but with the acceptance of Hess' theory of Sea floor Spreading there is little room for this objection now*).

- Equatorial and Polar diameters of the Earth are insignificantly different. Thus many believe that it is unlikely that equatorial convective current would be so strong to break the equatorial continent of Pangaea.

## Plate Tectonics and Mountain Building

It is obvious from the foregoing sections that despite severe criticisms of **Wegener**'s views on continental mobility the concept particularly after **Holmes'** support continued to generate positive interest amongst the Earth Scientists. In the meantime, evidences from present Southern Hemisphere accumulated to support the existence of Gondwana land, a breakaway part of the super continent of Pangaea that is suggested to have existed before 250 million years. Gondwana land at that time is believed to be near South Pole and which gradually broke in different segments of present India, Australia, South America, Africa and Antarctica (see chapter 4). It was also found that most of the oceanic crusts correspond with the Cretaceous period i.e. most of them are not older than 135 million years in age. These findings along with other evidences collected after 2nd World War led **Herman Hess** of Princeton University and **R.S. Dietz** of U.S. Coast and Geodetic Survey to propose the theory of Sea Floor Spreading in 1960. **Walter Elsassar**, a colleague of Hess also was convinced that the ocean floors were not an inert entity. They were very mobile and active. Mobility of the ocean floors i.e. a part of the lithosphere,

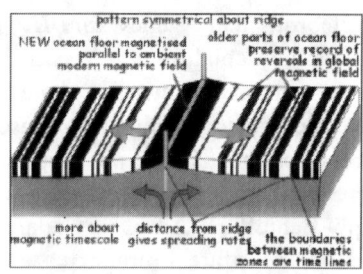
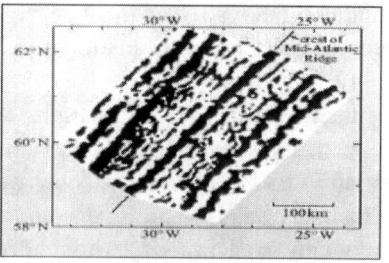

(https://www.see.leeds.ac.uk/structure/dynamicearth/index.htm) from **Bruchfield** et.al
(1982:175)

**Fig. 9:4:a: Oceanic Ridge and**
**Magnetic pattern**

**Fig. 9:4:b: Magnetic anomaly in**
**The Atlantic south of Iceland**

was attributed to convection currents (*suggested earlier by Holmes*) of
hot rising magma from the mantle. It was also suggested that where the
rate of rising magma from the upper mantle was slow it created rifts
along the Mid-Oceanic Ridges where they oozed and cooled in their median
valleys to form lithosphere. This was also suggested to be the zone where
convection current diverged a process that would force the two sides of
the oceanic ridges to move away from each other. The proposition also
found support in 1963 from **F.Vine** and **D. Mathews** in their study of
magnetic anomaly in rocks on both sides of mid- Atlantic ridge south of
Iceland. They also found juxtaposed angles of magnetic field on the two
sides of the ridge. Later same pattern was found in almost all the oceans.
According to them this could happen only when the ocean crust diverged
at the same rate away from mid- -Oceanic Ridge where molten rocks up
welled, cooled and magnetized in their median valleys to form new crustal
lithosphere. Later it was found that all mid-oceanic ridges did not have
median valley. It was discovered that no median valley was formed where
rate of spread of oceanic ridge was faster as in the East Pacific Ridge
system. (**Burchfiel** et.al: 1982:175). Yet magnetic anomalies on both
sides of the ridge were found to be symmetrical. These anomalies have
been found to be different from reversal of polarity of the Earth's magnetic
field in the sense that they are found with greater frequency. In case of
reversal of the Earth's magnetic field its polarity changes in about 700,000
years. It is expected to be so recorded in the Earth's palaeo magnetic
history.

All the studies since 1965 though appear to have confirmed creation
of lithosphere along the mid-oceanic ridges the size of the Earth has not
been found to increase. It has, therefore, been concluded that the crust is
being consumed by almost the same proportion somewhere. Independent

studies of **Kiyoo** *Wadati* *of* Japan Meteorological Agency, *and* *Hugo*
**Benioff** of California Institute of Technology conducted way back in
1949 had already shown that where the continental margins came in
contact with the oceanic margins ocean beds being thinner and denser
dipped almost at an angle of 450 under the continental lithosphere up to a
depth of 650- 700 km. The zones where the margins dip are now known
either as Wadati- Benioff zone or simply Benioff zone. It is this balance
between creation of lithosphere at some plate boundaries and
corresponding subduction on some other boundaries that does not allow
any significant volumetric change in Earth's dimension.

Combination of all these studies and their findings led to the
formulation of **New Global Tectonics** now popularly known as **Plate
Tectonics.** The theory is constructed to explain the mechanism of evolution,
nature and motion as well as resultant reactions of plates of which the
Earth's crustal lithosphere is believed to have been formed. It is
noteworthy that most of the plates are composed both of oceanic and
continental crusts. Only exception is provided by the Pacific plate which
is composed entirely of oceanic crust. It is a theory that attempts to
explain the formation and distribution of geological phenomena on the
Earth on the basis of movement and interaction of different plates and
their segments. The term 'Plate' was first used in 1965 by **John Tuzo
Wilson** of Canada. He used the term in the wake of defining    e*Transform
Faults'*. He also found that the 'spreading axis along the mid-oceanic
ridges, magnetic anomalies and topography of the ridges' were 'inherently
discontinuous'. He considered it to be a necessary corollary of the crustal
movements. He also stated that the discontinuous spreading axes were
connected with the nearby axis by a right angled fracture zone. He named
such fracture zones **'transform fault'**. These understandings of sea
floor morphology have been found to explain seismicity, volcanism,
continental drift, and mountain building besides distribution of many other
geological phenomena. However, it is **W.J. Morgan** of Princeton
University who is generally accredited with the formulation of the Plate
Tectonics hypothesis in 1967. But it must also be known that though
independently but almost simultaneously, British geophysicists **D.P.
McKenzie** and **R.L. Parker** reached almost identical conclusion. In fact
it was they who applied Euler's theorem to explain variability of mobility
and directions of different crustal plates. They opined that plates' movement
were controlled by their varied but independent poles inducing differential
movements in them. All oceanic ridges according to them formed 'great
circles'whereas the transform faults formed 'small circles'. The plates

tended to move along these circles. Under such dispensations different plates of lithosphere will have different poles of movement bringing plates either closer to each other or away from each other. Reconstruction of land masses in geological pasts suggests that the continental blocks have come closer and gone farther from each other many times in past (see chapter 4). Thus, the formulation of Plate Tectonics theory in its present form may equally be attributed to Wilson, McKenzie and Parker besides Morgan.

Based on new evidences collected after 2nd World War and additional inputs obtained from the study of thickness of sea floors sediments and magnetic polarity since 1960s 7 major plates have generally been identified. They are named after major landform features found over them. They are, thus, named as (i) **North American**, (ii) **Pacific**, (iii) **Eurasian**, (iv) **African,** (v) **Indo-Australian** often considered to be two independent plates having Australia and Indian sub continent, (vi) **South American**, and (vii) **Antarctic**. Most of these plates are combination of continental as well as oceanic crusts. Only exception is provided by the Pacific Plate which is composed entirely of oceanic crust. Besides these plates many minor plates have also been identified and which might have been broken away from some bigger plates in earth's geological past. Amongst many others some of them are: (a) **Arabian plate**, (b) **Philippine plate** (c) Cocos plate, (d) **Caribbean plate**, (e) **Nazca plate**, (f) **Scotia plate**, (g) **Juan de Fuca plate**. Like the big plates some of them like Nazca and Cocos plates are composed mostly of oceanic crust whereas Arabian plate is entirely continental. Some of these like that of the Philippines plate prominently reflect volcanic features (see fig. 9:2).

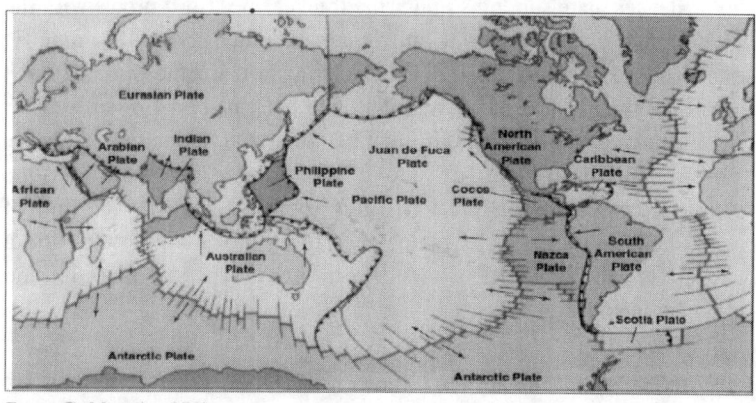

From **Gabler** (p. 378)

*Fig. 9.5: Most commonly identified Plates*

Identification of plates necessarily involves understanding of their interaction with neighbouring plates. It has now been established that these plates inter act in three ways

(i) moving away from each other;

(ii) sliding past each other

(iii) coming closer to each other.

In the light of these possibilities plate are involved respectively (1) in creation of lithosphere as found along the spreading axes, (2) sliding past each other horizontally along transform faults, (3) drowning of one plate under the other.

(1) In the case of (1) there may be continual formation of new lithosphere under the oceans due to upwelling of magma from the upper mantle and its cooling along mid-oceanic ridges. They result in formation of new lithosphere dominated by rocks like gabbro at greater depth and basalt at lesser depth. These boundaries where new crusts are formed are known as Spreading Boundaries. The zones where spreading take place are known as **Constructive** or **Divergent margins.**,

(2) It is also possible that plates simply pass by each other along some deep seated faults.

They would not be involved either in creation or loss of materials of which plates are composed of. The margins where it happens are called **Shear** or **conservative margins,** and

(3) if plate boundaries come closer to each other thinner and denser plate subducts under the thicker ones till it mixes with the hotter mantle material and under goes density and volumetric change. Almost all the oceanic trenches are the sites of such activities. Such plate margins have been named **Consuming** or **Subduction margins**. It must be noted that the plate boundaries demarcate places of above mentioned three interactions whereas the term plate margins is used to denote the end of the plate. A single plate, therefore, has margin and not a boundary. Fig. 9:3 shows the different types of margins and boundaries

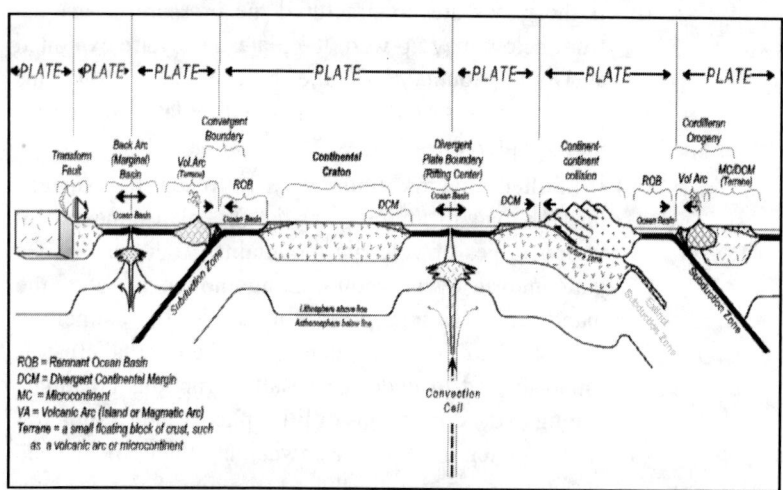

After csmres.jmu.edu/geollab/vageol/vahist/plates.html

***Fig. 9:6: Plate Boundaries and Inter plate Relationships***

As there is a general agreement in respect of mode of interaction between crustal plates the proponents of the theory have been able to outline the evolution of tectonic mountains as follows:

- Mountain ranges are formed along the active belts of crustal plates.

- Active mountain-making belts are narrow zones that are usually found along the plate margins.

- There exists a definite relationship between seismicity and mountain building. In other words, formation of the mountains is impacted by the operation of convective shells that operate in the lower lithosphere and upper mantle and which are involved in creation of crust or their subduction.

- Mountains are formed either due to extensional or compressional activities induced by sub terranean convective currents.

    (a) Extensional tectonic activities lead to thinning of crust as along the spreading margins. Where the rising columns of convective currents diverge crustal plates are exposed to tensional forces leading to fracture or vein formation in the affected crust. It leads to the formation of three kinds of mountains.

    (i) If plate is fractured parts of it may be pushed up to form 'Block Mountains'.

(ii) If the plates are not fractured the pressure of magma from below may up warp the plates. They are exhibited as Dome Mountains on the surface. Such mountains are not necessarily associated with plate boundaries (see chapter 8).

(iii) If molten magma finds way to reach the surface through one or more veins in the plates they form various types of volcanic mountains. Additionally, if a plate moves over a non-moving molten mass of the mantle called *mantle plumes* it may lead to its melting at the base of the upper mantle and the oceanic crust. It found to lead to undersea basaltic eruptions creating / adding to the dimensions of lithosphere. Such a situation is encountered in the Haawaaiian chain of islands evolving over the northwest moving Pacific plate. Areas, where new plates are being created due to upwelling of magmatic materials as in case of spreading margins and subducting margins, are known as *hot spots.* Hot spots are found both under continental as well as oceanic plates. They are considered to be an important instrument in the growth of plates.

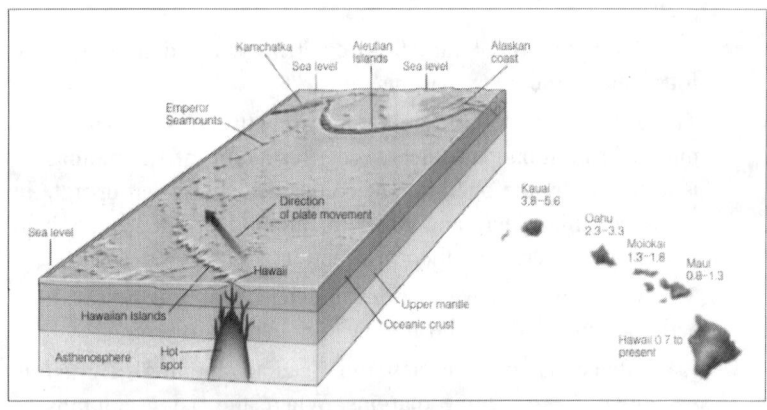

From **Gabler** (2007:386)

***Fig. 9.7: Hot Spot and plate characteristics along Hawaii islands***

(b) Compression tectonic activities, on the other hand, are involved in crumpling and shortening of crusts. This may happen in either of the two ways (a) One plate subducts under the other (non-collisional shortening of the crust), or

(b) two plate boundaries collide with each other (collisional shortening of the crust). As a result of long compressive stresses, thus, Volcanic and Fold Mountains are formed. They are very distinctly distributed and are found to exist in following three types of zones:

(a) Between two island arcs i.e between oceanic-oceanic plate margins;

(b) Between an arc and a continent i.e. along oceanic-continental margins; and

(c) Between two continental plates i.e along continent-continent margins.

In case of (i) and (ii) fold mountains along with volcanoes are created in the coastal parts. They are necessarily associated with adjacent oceanic trenches. The best example is found in circum Pacific belt where they exhibit themselves in a series of arcs from Ryukyu in western Pacific through Honshu and Kurile to Aleutian Islands in north eastern Pacific. Presence of the western mountain ranges and their association with chain of volcanoes along the western costs of North and South America is also attributed to convergence of ocean-continent margins.

Fold Mountains have also been formed at convergent boundaries of two continental plates having some times in geological past geosynclines in between. They are exemplified by the Himalayas and the Alps between the Indian and Eurasian and African and Eurasian plates respectively and which occupy the space of Mesozoic Tethys Geosynclines.

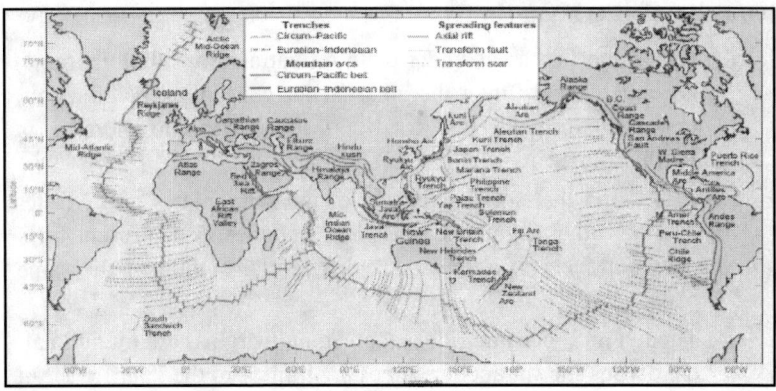

From **Strahlar** (p. 226)

*Fig. 9.8: Distribution of plates and mountains*

Evolution and vertical dimension of mountains attain isostatic equilibrium by adjusting **gravity anomaly** of rocks under their roots.

Following figure shows the areas of recent episodes of mountain formation in different parts of the world.

## Merits of the Hypothesis

Plate Tectonics has taken into account the logical postulates of earlier hypotheses and is an integrative theory. It is found to explain most of the tectonic features of the Earth.

*   It provides a rational explanation for the present distribution of continents and oceans.
*   Tectonic activities associated with the formation of volcanoes and distributions of earthquakes are well explained.
*   It logically explains the formation and distribution of coastal as well as continental mountain systems.
*   It also is able to explain igneous intrusions and their impact on crustal lithosphere particularly in areas of mountain building.
*   The theory also addresses the distribution of negative landforms of the ocean basins.

It may, however, be noted that despite its popularity and wide acceptance the theory of plate tectonics is yet to explain many of the apparent facts. Unless these facts are explained there are certain limitations on its universal acceptance.

## Demerits of the Hypothesis

*   It is found that areas under plate formation (spreading margins) are much bigger than the areas under consuming (subduction) margins. Normally this condition should lead to increase in Earth's dimension.
*   It is found that some segments of the same plates move faster than the other segments. The plate tectonics in its present form does not seem to address this differential motion of the same plate.
*   Though mid- oceanic ridges (spreading margins) are found in all the oceans the subduction zones are limited to the Pacific Ocean only. Why it is so? The theory does not explain.
*   Why Benioff zones is not found along the North American-Pacific plate boundary.

o   Same plate is found some times moving in two different directions. The theory seems to overlook the problem. It may be caused by rising and diverging convective currents under the continental plates. But the theory suggests such currents only under the oceanic crusts.

■   The theory in its present form explains the formation of tertiary mountains but fails to address the formation of older mid-plate mountain ranges found over many table lands in different parts of the world.

Despite its above mentioned short comings the theory is still able to provide most plausible mechanism for the formation of different crustal features. Integration of more information forthcoming from the study of the nature of interior composition of the Earth and their analysis may help in providing answers to these objections.

## Review Questions

1.   Identify the major characteristics of global orogenic belts.
2.   Describe the major phases in the evolution of geosynclines. How are they associated with mountain building?
3.   Evaluate the mechanism of mountain building as suggested by Kober.
4.   Assess the validity of propositions made either by Jaffereys or Joly or Daly or Holmes.
5.   Write short notes on the following:-
6.   Concept of Hinterland and foreland,
7.   Continent- ocean convergence and Island arcs
8.   Benioff zone
9.   Magnetic anomaly
10.  Fire girdle of the Pacific
11.  Hot spots
12.  Gravity anomaly
13.  What is a lithospheric plate? Identify three types of plate boundaries and their functions. Give suitable examples.
14.  What differentiates the continental crust, oceanic crust, lithosphere, and Asthenosphere?
15.  Evaluate the evidences and mechanism provided by Wegener to support his theory of continental drift. What evidences now have been found to support his theory?
16.  Name the types of plate boundaries responsible for the evolution of the western mountain system of South America, Alpo-Himalayan system of mountains, creation of the San Andreas Fault, in Iceland, and near the Himalayas?
17.  Bring out the significance of the Hawaiian Islands with reference to plate tectonics theory?

18. Describe the process of subduction as it occurs at a converging boundary of continental and oceanic lithospheric plates. How is the continental margin extended? How is subduction related to volcanic activity?

19. What are *transform faults?* Where do they occur? How are they associated with seismic activities?

20. Name the six great lithospheric plates. Identify an example of a spreading boundary by general geographic location and the plates involved. Do the same for a converging boundary.

21. How are island arcs formed? What type of plate collisions is involved?

22. What is meant by the term *arc-continent collision*? Describe how it occurs.

23. How is the principle of convection thought to be related to plate tectonic motions? What role might gravity play in the motion of lithospheric plates?

## References

1. Bullard, E.C.; Everett, Jim and Smith Allan (2008): Genesis of a Geophysical Icon: The Bullard.

2. Everette and Smith Reconstruction the Circum-Atlantic Continents; Earth Science History, Vol. 27, No. 1; pp. 1-12.

3. Chander, Ramesh (1999): *'Wegener and his Theory of Continental Drift'*; *Resonance*; Dept. Of Earth Sciences, University of Roorkee; pp. 24-41.

4. Dayal, P. (1990): A Text Book of Geomorphology; Shukla Book Depot, Patna.

5. Letsch Dominik (2015): 'R.A. Daly's early model of seafloor generation 40 years before the Vine- Matthews hypothesis: an outstanding theoretical achievement inspired by field work on St. Helena in 1921-1922'; *Canadian Journal of Earth Sciences,* Vol. 52 *issue* 10 pp. 893-902.

6. Smith, A.G. and Hallam, A. (1970): The fit of the southern continents; Nature' 225: 139-144.

7. Sproll. W.P and Dietz, R.S (1969): 'Morphological Continental Drift Fit of Australia and Antarctica'; Nature, 222:345-348. https://www.see.leeds.ac.uk/structure/dynamicearth/index.htm

8. Strahlar, A.N. (2011): Introducing Physical Geography, John Wiley & Sons

## Note

1. Suess believed that only one land mass could move and the other would remain stationary. He identified the moving land mass as 'Backland' and stationary one as 'Foreland'; Argand in tune with the later postulates of Wagener's continental drift theory discarded the concept of stationary land blocks. He believed in the differential motion of the two blocks in the same direction. He named faster moving block as 'hinterland' and the slower moving one as 'foreland'.

# Chapter - 10

# Earthquakes and Volcanoes

In earlier chapters attempt has been made to understand various visible and verifiable characteristics of our ever changing planet. In present chapter an attempt is made to understand the turbulences within the interior of the Earth and which are found to have been shaping its surface features since its consolidation and have been impacting its existing dispensations. This chapter, therefore, is devoted to the understanding of the causes, distribution and impact of

      Earthquakes, and

      Volcanoes

## Earthquakes

The Earth, as has been discussed in forgoing chapters, is a dynamic planet and its dispensations are perpetually impacted by endogenic and exogenetic forces. Earthquakes are one of the endogenetically produced such forces that have been responsible for shaping the landforms since the crust started being solidified some 4.2 years ago. They are the manifestations of the Earth's ever changing energy balance in its interior (see chapter-8). Few of the earthquakes though may take place under the impact of extraterrestrial objects like meteors most earthquakes are generally caused either by tectonic adjustments in segments of lithosphere (*tectonic earthquakes*) or by magmatic eruptions through volcanoes (*volcanic earthquakes*). In all the above mentioned cases a sudden disturbance and adjustment of earth forming materials take place. The earthquake, thus, may be defined as a *sudden shaking and trembling of the* Earth's surface consequent upon release of accumulated pressure and accordant rock *adjustments in lithosphere*. It has to be noted that release of pressure may take place along active fault planes as well as in areas of magmatic movements within active volcanoes. These activities take place either due to collision of plates or due to subduction of one plate under the other. Earthquakes are accordingly classified as **collisional** or **noncollisional** respectively. In both the cases seismic waves are generated. Seismic waves, thus generated are transmitted radially from the point of their origin through the Earth. It is more like ripples created by pebble thrown in a water pool (see figure 10:1).

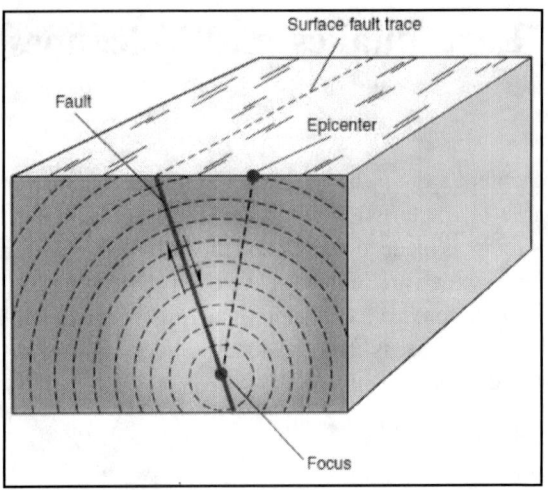

From Gabler: 413

**Fig. 10:1: Mode of transmission**

Earthquakes are recorded in any part of the globe with the help of an instrument called **seismograph**. The point of release of stress leading to earthquakes in the interior parts of the earth is known as the **focus** or **hypocenter** and the point just vertically above it on the surface is called **epicentre** of the earthquake (see chapter 5 as well as fig, 10.1). It is noteworthy that almost all the earthquakes are preceded by **foreshocks** and followed by **aftershocks** of varying magnitude. Though generally lesser than the main shocks they, particularly the aftershocks, are by no means less destructive in the affected regions. Magnitude of these shock waves has generally been measured on different types of scales since middle of 19th century. M. S. De Rossi and F.A. **Forel** developed a ten point descriptive scale in 1883 to ascertain the intensity of earthquake. It was based on observation and quantum of damage brought about by the earthquakes. It ranged from micro seismic tremor recorded only on a single seismograph to extremely high intensity tremors recorded on a number of seismographs and that brought about great disasters, disruption of the earth strata and fissure eruptions. Their ten point scale was denoted by Roman numerical from I to X. The scale was further modified by **Giuseppe Mercalli**, an Italian volcanologist between 1884 and 1906. He attempted to quantify the effects of earthquake on Earth's surface, human beings, animal kingdom as well as man-made structures. Mercalli with the aid of his compatriot physicist **Adolfo Cancani** and German geophysicist **August Heinrich Sieberg** developed a twelve point earthquake scale from I (not felt) to XII (total destruction). Charles F. Richter, an American seismologist of German parentage, further modified this intensity scale. It is known as **Modified Mercalli Intensity** Scale (**MMI**). By 1935, he, however, developed another ten point earthquake magnitude scale in collaboration with **Beno Gutenberg.** They together made an attempt to correlate the seismic intensity with that of release of energy. The scale, however, is popularly known as Richter scale.

In this method each whole number represents 10 times greater ground motion than the previous one. It has been calculated that each number on the scale releases 31.5 times more energy than the preceding one. The scale, thus, is able to suggest local displacement quantitavely. It is due to this reason that it is reflected on earthquake charts as $M^L$.

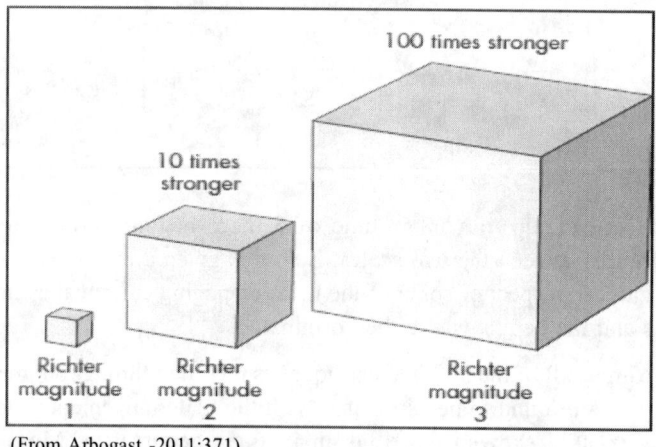

(From Arbogast -2011:371)

***Fig. 10:2: Relationship between motion and Enetrgy Release on Richter Scale***

Richter scale, however, has been replaced by a more accurate **Moment Magnitude Scale (MMS)** since 1993. It is reflected on earthquake charts as MW. This scale was developed by T. Hanks and H. Karmori in 1979. The scale uses more variables in respect of release of energy. It involves measurement of the magnitude of fault slippage, its nature and size as well as the size of the affected surface and the nature and strength of the rocks that rupture due to seismic activities. It is believed to be more accurate than the Richter scale which is concerned with the measurement of magnitude of seismic waves and is discerned as soon as data from different seismograph stations are collected. On the other hand, Moment Magnitude Scale is based on observed effects of the earthquakes and requires information from the affected areas to evaluate the extent of damage. Below is given a comparison of three most widely used scales for the measurement of magnitude of the seismic effects.

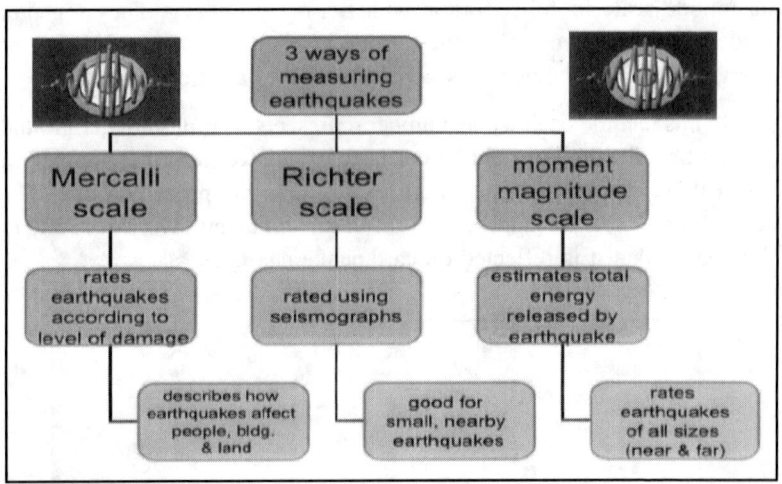

However, up to a magnitude of 5 there is found to be no much difference between the two scales. Below is given a comparison of both the scales in respect of some of the most damaging earthquakes in recent times and the depths where they originated.

Almost all of the 500,000 earthquakes recorded throughout the world in a year are found to be associated with the plate movements. However there exists a skewed distribution in their occurrence. Most of the earthquakes (about 80% of them) are found to be associated with noncollisional subduction plates. They are most often associated with volcanic activities along the **Pacific Rim.** It is this characteristic that is reflected in the nomenclature *"Pacific Ring of Fire"* for the region. The region accounts for over 75% of the active and dormant volcanoes on global level. Rest of the volcano related earthquakes are found along spreading margins of the plates and associated transform faults (to be discussed later in the section). Tectonic earthquakes, on the other hand,

## Table: 10:1: Most damaging earthquakes since 2004

| Date & Year | Location | Magnitude of Damage Deaths /Property loss | Richter scale ($M^L$) | Moment Magnitude Scale ($M^w$) | Modified Mercalli intensity scale ($MM^I$) | Depth of Focus (in Km.) |
|---|---|---|---|---|---|---|
| 26.12.2004 | Indonesia, Indian Ocean | >227,900;?; $10 bn. USD | 9.1 | 9.3 | IX | 30 |
| 08.10.2005 | Kashmir | > 83,000; All villages with mud brick home flattened; economic loss-? | ? | 7.6 | VIII | 15 |
| 12.05.2008 | Chengdu (China) | >90,000; 5 mill. Homeless; economic loss- $122 bn. USD | 8 | 7.9 | XI | 19 |
| 12.01.2010 | Port-au-Prince (Haiti) | >316,000; 1.5 mill. Homeless; economic loss-? | 7.3 | 7.0 | XI | 13 |
| 27.02.2010 | Maule (Chile) | 700/.5 mill. Homeless; economic loss-? | 7.1 | 8.8 | VI-VIII | 35 |
| 11.03.2011 | N.E. coast of Japan | >18,000; 2668 missing; economic loss- $360 billion USD | 8.9 | 9.0 | IX | 42 |
| 24.09.2012 | Baluchistan (Pakistan) | >825; 350 home village destroyed | ? | 7.7 | VII | 15 |
| 03.08.2014 | Yunnan (China) | 700; 12000 homes destroyed | 6.5 | 6.2 | VIII | 12 |
| 25.04.2015 | Nepal, Tibet, Bangladesh | >8000; 8 mill. People affected; | 8.1 | 7.8 | VIII | 15 |
| 16.04.2016 | Ecuador | >670; | ? | 7.8 | VIII | 20.6 |
| 29.07.2016 | Nn Mariana Island (USA) | ? | ? | 7.7 | ? | 212.4 |
| 17.12.2016 | Papua New Guinea | ? | ? | 7.9 | V TO X | 103.2 |
| 08.09.2017 | Chipas (Mexico) | 41,000 houses damaged, 98 dead, 300+injured | | 8.2 | IX | 47 |
| 19.08.2018 | Fiji Region | ? | ? | 8.2 | ? | 563.4 |
| 21.8.2018 | Río Caribe Venezuela) | ? | ? | 7.3 | ? | 154.3 |

*Source:* USA TODAY research and the U.S. Geological Survey.

are caused by relative and differential motion of adjoining plates and their accordant structural adjustments. Stress generated by continual friction leads to the formation of fractures (faults) and displacement of earth materials between adjacent plates or under them. These fractures, depending on the nature of the plates and intensity of stress, may be shallow or deep rooted. It, accordingly, transmits seismic waves through the lithosphere. Despite the fact that most of such earthquakes are generated along the plate boundaries and active transform faults few of them may have mid-plate origin where small faulting may take place due to certain internal stresses.

Tectonic Earthquakes are generally classified on basis of the depth of focus which may occur from a few kilometres up to 700 kilometres of depth. They are classified in three groups as:

- **Shallow earthquakes** having a focus generally above 70 Km. of depth;

- **Intermediate focus earthquakes** having a focus depth ranging between 70 to 300 Km.; and

- **Deep focus earthquakes** ranging in depth from 300 to 700 Km.

In this respect it is noteworthy that the tectonic plates being composed of brittle and relatively cooler rock materials and descending down in hot mantle are the only parts which can store elastic energy till a depth where a temperature of 300° C is reached. This elastic energy is released when fault rupture takes place. Magnitude of tectonic earthquakes is, thus, related to the generated stress which is a dependent variable to the length and breadth of the faults that develop. Bigger the faulted area greater would be the magnitude of the earthquake. Most of the big ruptures of this kind have been observed at different times in subduction zones along Chile coast, Alaska coast and Sumatra. On the other hand, ruptures associated with some of the biggest strike-slip faults like those of San Andreas Fault of the USA, Anatolian Fault in Turkey and Denali Fault in Alaska are much smaller in dimension. The size of rupture is still smaller along the normal faults occurring in the continental plates. Size and depth of rupture determines the spatial dimension of the impact of the earthquake. Volcanic Earthquakes, on the other hand, are essentially associated with dynamic igneous activities under the lithosphere. Eruption of magma, its viscosity and it's cooling in sub surface and over surface create conditions for earthquake. They are the results of natural processes and are responsible for the formation of intrusive and extrusive rocks. Magma and subsurface rock materials (**tephra**) along with gases find their way through some openings. These openings through the earth's lithosphere are called **vents** or **dikes** (see classification of igneous rocks in chapter 7). Characteristic volcanic landforms depend primarily on (a) chemical composition of the magmas which is related to the pockets it emerges from; (b) its propensity to flow which is determined by its viscosity that ranges from very low (very fluid and rapidly moving) to very high (thick and very slow moving) and (c) the nature of pyroclasts (obtained from Greek words pyros = fire; and clasts = broken) determined by their size. Generally three categories of pyroclasts are identified on the basis of their size. They are called (i) *Bombs* (more than 64 mm in size), (ii) *Lapilli* (size 2 to 64 mm) and (iii) *ash* (size less than 2 mm.) respectively. Collectively they are known as *tephra*. Initial landforms in volcanic regions are formed by these magmatic materials.

Volcanism most potently operates along the spreading and subducting plate margins. It varies in size and character and produce diverse type of subsurface (Batholiths, laccoliths Sills etc.) and surface landforms like shields, cones and volcanic mountains (see chapter 7). Three hierarchical processes are recognized that produce volcanoes and associated landforms. (i) Liquefaction under certain depth (a depth below the surface where more than 300° C temperature is encountered) of the Earth. Volume expansion creates pressure enabling magma to rise through certain opening (conduits). (ii) Mixing of magma with subsurface gases and water vapour. It hastens the rate of its mixing and provides explosive energy to magma. (iii) And ejection of magmatic materials in the form of lava and solid pyroclasts when the strength of surface rocks is surpassed. They together produce different kinds of landforms. It is mainly because the ejected materials differ in their temperature and chemical composition. These materials are classified on the basis of size of crystals and minerals they are made of. They are broadly classified as mafic and sialic (see igneous rocks in chapter 7). Mafic magmas having greater proportion of iron and high melting temperature are less viscous and flow easily. Sialic magmas, on the other hand, have lower melting temperature and are more viscous. They do not flow easily. During their upward journey from mantle to the surface magmas undergo qualitative changes and erupt accordingly. If not retained by the magmas decrease in pressure leads to release of expanding gases and tephra on the surface. In absence of any opening through to surface gases aided by pyroclasts of different sizes erupt explosively to create openings. Depending on the number of conduits, nature of mineral composition and viscosity of lava and size of pyroclasts different eruptions vary in their intensity and capacity to modify landforms. And though no two volcanic eruptions are identical on the basis of broad similarity of the mode of eruption volcanoes may generally be grouped in two categories. They are formed due either to *Explosive eruption*, or *Effusive eruption.*

**Explosive eruption**: Explosive eruptions are more common in Volcanoes with central dikes and associated off shoots. Such eruptions involve magma with high viscosity and high proportion of gas content. Explosiveness of gas content breaks and ejects the viscose magma as pyroclastic matters (tephra) which tend to accumulate layer over layers. They are, thus, responsible for the creation of what are known as **strato** or **composite volcanoes**. Very often, however, volcanoes expel only loose and grainy cinders of basalt and Andesite without much trace of lava. They form most simple and common type of volcanoes known as **cinder cones**. Cinder cones are the volcanoes which have small horizontal and vertical dimensions.

 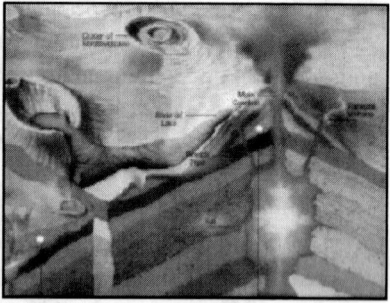

Courtesy Britannica Illustrated Science Vol. 3 (p. 28)

    *Fig. 10.3a: Cinder Cone*        *Fig. 10.3: Composite volcanoes*

They are generally found to be 1 to 2 kilometres in width with an altitude of about 300 metres. Based on the intensity of different associated activities five types of explosive volcanoes are generally recognized. They are named as (i) *Strombolian*, (ii) *Vulcanian*, (iii) *Visuvian*, and (iv) **Pelean**. Eruptions of these types differ from each other in respect of manner and the type of materials they eject. It, however, is possible that the same volcano has different types of eruption at different times.

**Strombolian Type**: Named after the volcano Stromboli in Sicily this type of eruption may continue episodically for centuries. They are formed by the rise of magma containing gas bubbles in it. These gas bubbles are

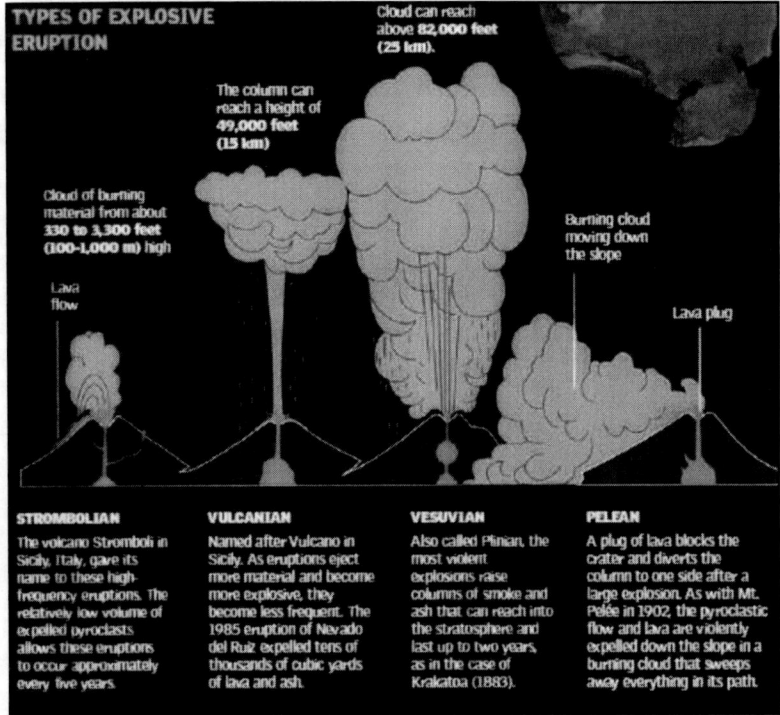

(From Britannica Illustrated Science Library, Vol. 3: 20)

*Fig. 10.3: Types of explosive eruption*

known as **gas slugs.** These gas slugs rise with relatively more viscous basaltic magma as compared to that of the Hawaiian type. They burst with difference in pressure and explode while moving upwards towards the surface in nearly continuous manner. Ejected materials may be discerned from a distance due to magmatic glow. Eruption may rise up to hundreds of meters above the surface containing some tephra. This kind of eruption is very often associated with lava lakes which generally develop in conduits of volcanoes.

**Vulcanian Type:** A Vulcanian type of eruption named after Vulcano in Sicily is characterized by sporadically short duration but repetitive violent explosion of very dense magma. The explosion is caused by quick cooling of magma and fragmentation of solid lava (lava plugs) in the conduits of volcanoes. Explosion may also be caused by rupture of accumulated solidified lava over vent known popularly as **lava dome.** It takes place when gas under the blockages accumulate in great quantity and exert pressure more than the strength of the solid magmatic materials that block its passage. These explosions containing tephra may rise many kilometres above the surface of the Earth. The eruption may continue for few days to years.

**Visuvian Type:** Visuvian type of eruption, witnessed and recorded by Pliny, a Roman historian in 79 A.D., is also known as Plinian type. They are known to be the most violent of the eruptions and very destructive. These eruptions are characterized by the fragmentation of very viscous magma and release of great amount of gases there in. Initially magma is erupted from the lateral cracks of the volcanoes enabling the gases to accumulate in main vent. It is only after some pressure is released (due to emission of magma from the lateral cracks) that eruption starts from the main vent. Enormous amount of energy released in this manner forces the volcanic materials to attain a very high velocity and send the ejected materials very high and wide in the atmosphere. Such eruptions have been found many times to rip apart the top portions of volcanoes. Columns of erupted materials that include gases and ash attain a shape similar to mushroom and may reach a height of 50 Km. above the surface. Ash from such eruptions may reach thousands of miles horizontally from its source. The eruption of this kind is generally so strong that magma chambers are emptied for some time and volcanoes become dormant before sufficient energy is stored for next eruption.

**Pelean Type**: Named after Mount Pelee in West Indian island of Martinque this type of volcanic eruption is very similar to that of the Vesuvian type. It, however, is characterized by greater viscosity of lava and small amount of tephra. It also differs in the mode of eruption in which mixture of lava and gas is not thrown upward. Instead they move rapidly down slope. This happens mostly because more viscose lava coagulates rapidly in the main vent obstructing its escape. As a result, entrapped gases are released through the lateral vents followed by gas charged lava admixed with fragmented materials and ash.

**Effusive Eruption:** Effusive eruption refers to outpouring of mafic (basaltic) magma over a large area through one or more fissures. It is characterized by a continuous and steady flow of degassed lava. Degassing of magma depends on its permeability. Other factor that influences eruption behaviour is the rate of ascent of magma. Ash content in this kind of eruption is almost nonexistent. Such eruptions are known as **Hawaiian type** of eruptions. These types of eruption are mostly found along the subduction belts of lithosphere. Gases and lava in these types of eruptions are ejected separately. Tephra in these types of eruptions is almost have a lava thickness between 1000 and 1300 metres. Basalt plateau of Iceland is also believed to have evolved due to this kind of eruptions. They all stand testimony to effusive eruption in geological past covering periods between Cretaceous in India to Miocene in Columbia and Iceland.

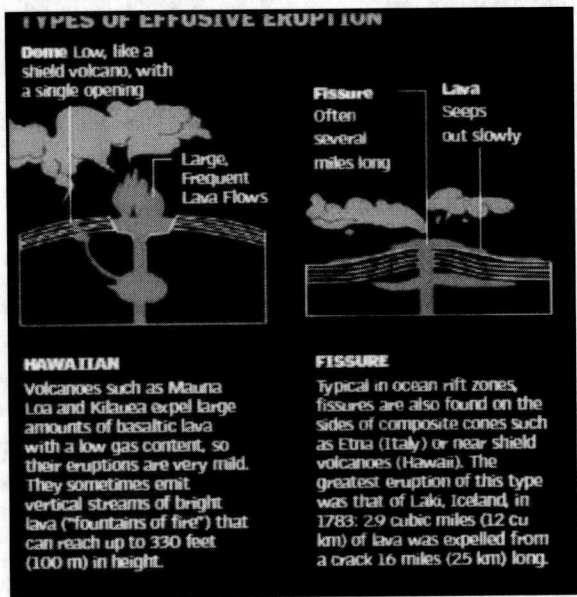

From Britannica Illustrated Science Library, Vol. 3: 20.

*Fig. 10.4: Types of Effusive Eruption*

nonexistent. These eruptions are generally very mild releasing very low amount of gases and very large amount of thin basaltic lava which are spread far and wide with relatively low elevation. Mauna Loa and Kilauea of Hawaii Islands are the representative volcanoes of this type of eruptions.

Effusive eruptions releasing more degassed fluid lava are also responsible for spreading ejected materials far and wide with relatively low vertical elevation. They are, therefore, also responsible for the evolution of **Shield Volcanoes**. Basalt plateaus of Columbia and Iceland are believed to have been formed due to these kinds of eruptions. Columns of lava in these types of eruption may rise up to a height of 100 meters. Deccan lava plateau of India, Basalt plateau of Columbia in U.S.A. and Plateau of Iceland are some representative landforms developed as shield volcanoes. They have left behind an area of about 3 lakh Km2 in India with thickness of lava ranging between 700 and 1600 meters. Similarly, Columbian basalt plateau that roughly covers an area of about 1.5 lakh Km2 is suggested to

**Other Classification of Volcanoes**

One of the ways to classify Volcanoes is the frequency of their activity. On this basis volcanoes are categorized in three groups as (i) *Active,* (ii) *Dormant,* and (iii) *Extinct.*

(i) **Active Volcanoes:** Active volcanoes technically refer to those volcanoes which have erupted at regular intervals in known history of mankind. Some volcanologists consider volcanoes to be active even if they have erupted only once in last 10,000 years. By this definition about 1500 volcanoes are considered to be active. Presently a little less than 600 volcanoes are believed to be active on global level. Though active volcanoes have been identified in many parts of the world about 75% of them are associated with the Pacific Ring of Fire. Some of the most destructive known active volcanoes are Mt.Mona Loa (Hawaii Islands), Mt. Etna (Sicily), Mt. Vesuvius (Italy), **Eyjafjallajökull (Iceland), Sakurajima (Japan),** Mount Merapi (Indonesia), **Mount Nyiragongo (Congo), Ulawun (Papua New Guinea),** Taal Volcano, (Philippines), Galeras (Colombia), and Santa María **(Guatemala). One active volcano known as Barren** Island volcano is also located in group of Andaman and Nicobar Islands of India.

(ii) **Dormant Volcanoes:** A dormant volcano is generally defined as one which had been not active for considerable period of time but has the potentiality to erupt. Definition is very subjective and includes volcanic activities from thousands of years to few decades interspersed with a considerable period of no eruption. Some of the dormant volcanoes as identified by volcanologists are given here with years of their **previous and recent eruptions or potentiality of eruption. They include** Mount St. Helens in U.S.A. (mid 1800>1980); Chaitén of Chile (7000 B.C > 2008); Eyjafjallajökull in Iceland ( 1800>2010); Mauna Kea, Hawaii Islands ( 2400 B.C. > Potential); Sete Cidades, Portugal (1880> Potential); Mount Teide, Canary Islands (1909>Potential); Mount Ararat, Turkey (1840> Potential); Solfatara, Italy (1158> presently releasing sulphurous gases); Mount Hood, U.S.A. (between 1814- 1836> 2006); Mount Fuji, Japan (1707> Potential), Popocatepetl, Mexico (1994>2000 > potential).

(iii) **Extinct Volcanoes:** It is very difficult to distinguish between dormant and extinct volcanoes. Generally, however, Volcanoes which have not been found to have erupted in last 10,000 years are suggested to be extinct. Such volcanoes are believed to have exhausted lava supply from their sources. They are also believed to have moved away from hot plumes present in mantle responsible for volcanic activities. On this basis many volcanoes world over are said to be extinct. Some of them include Mount Kilimanjaro (Tanzania), Mount Kulal (Kenya), Chimborazo (Ecuador), and volcanoes of Hawaiian Emperor seamount chain (North Pacific Ocean).

## Review Questions

1.  What is an *earthquake*, and how does it arise? What scale is used to describe the power of an earthquake? In what plate tectonic settings do earthquakes occur?

2.  Discuss earthquake processes, including why they happen, the nature of seismic waves, and how are they measured.

3.  Why are earthquakes typically associated with plate boundaries?

4.  What is a *volcano*? Why are volcanic eruptions environmental hazards?

5.  How is the global pattern of volcanic activity related to plate tectonics?

6.  Differentiate between a *stratovolcano and shield volcano*. What are their characteristic shapes, and how are they formed?

7.  What is a *hotspot*? What produces it? What landforms result from a hotspot on land? and in the ocean?

8.  Why are volcanoes associated with subduction zones?

9.  What are the differences between composite volcanoes and shield volcanoes?

10. What is the Pacific Ring of Fire and why does it have that name?

11. What is the relationship between an epicenter and the focus of an earthquake?

12. Differentiate among the Mercalli, moment magnitude, and amplitude (Richter) scales. How are these used to describe an earthquake? Why has the Richter scale been updated and modified?

13. What is a volcano? In general terms, describe some related features.

14. Where do you expect to find volcanic activity in the world? Justify your answer.

15. Compare effusive and explosive eruptions. Why are they different? What distinct landforms are produced by each type? Give examples of each.

16. What is the major difference between volcanism and solid tectonic activity? What characteristic structural and landform features develop due to these activities?

## References

1.  Arbogast, A.F. (2014): Discovering Physical Geography; John Wiley & Sons; New Jersey.

2.  Brittanica Illustrated Science Vol. 3 (2008): Volcanoes and Earthquake, Encyclopaedia Britannicca, Inc, New Delhi.

3. Burchfiel, B.Clark, Robert J. Foster, Edward A. Keller, Wilton N. Melhorn, Douglas G. Brookins, Leigh W. Mintz, Harold V. Thurman (1982): Physical Geology; Charles E. Merrill Publishing Co.; London.

4. Christopherson, Robert W. (2012): **Geosystems: an introduction to physical geography**, 8th ed. Prentice Hall, New Jersey, USA.

5. Dayal, P. (1990): A Text Book of Geomorphology; Shukla Book Depot; Patna.

6. Gabler, Robert E., Petersen, James F., Trapasso, L. Michael, Sack Dorothy (2009): Physical Geography, Ninth Edition; Brooks/Cole, Cengage Learning; Belmont, USA.

7. McKnight, Tom L. and Hess D. (2011): McKnight's Physical Geography: *A Landscape Appreciation* (11th edi); Pearson Education, Inc.; Delhi.

8. Strahler, A, (2011): Introducing Physical Geography; John Wiley & Sons; New Jersey. Wikipedia: Moment magnitude scale.

# Index